T0135535

Andreas Krug

Extension groups of tautological sheaves on Hilbert schemes of points on surfaces

λογος

Augsburger Schriften zur Mathematik, Physik und Informatik

Band 20

herausgegeben von:
Professor Dr. F. Pukelsheim
Professor Dr. W. Reif
Professor Dr. D. Vollhardt

Die Gutachter der Dissertation waren:

- Prof. Dr. Marc Nieper-Wißkirchen
- Prof. Dr. Samuel Boissière
- Prof. Dr. Marco Hien

Die mündliche Prüfung fand am 1.8.2012 statt.

Bibliografische Information der Deutschen Nationalbibliothek

Die Deutsche Nationalbibliothek verzeichnet diese Publikation in der Deutschen Nationalbibliografie; detaillierte bibliografische Daten sind im Internet über http://dnb.d-nb.de abrufbar.

ISBN 978-3-8325-3254-3
ISSN 1611-4256

Logos Verlag Berlin GmbH
Comeniushof, Gubener Str. 47,
10243 Berlin
Tel.: +49 030 42 85 10 90
Fax: +49 030 42 85 10 92
INTERNET: http://www.logos-verlag.de

0.1 Introduction

For every smooth quasi-projective surface X over $\mathbb{C}$ there is a series of associated higher dimensional smooth varieties namely the *Hilbert schemes of n points on X* for $n \in \mathbb{N}$. They are the fine moduli spaces $X^{[n]}$ of zero dimensional subschemes of length n of X. Thus, there is a universal family Ξ together with its projections

$$X \overset{\mathrm{pr}_X}{\leftarrow} \Xi \overset{\mathrm{pr}_{X^{[n]}}}{\to} X^{[n]}.$$

Using this, one can associate to every coherent sheaf F on X the so called *tautological sheaf* $F^{[n]}$ on each $X^{[n]}$ given by

$$F^{[n]} := \mathrm{pr}_{X^{[n]}*}\,\mathrm{pr}_X^*\,F\,.$$

More generally for any object of the bounded derived category $F^{\bullet} \in \mathrm{D}^b(X)$ using the Fourier–Mukai transform with kernel the structural sheaf of Ξ yields the *tautological object*

$$(F^{\bullet})^{[n]} := \Phi_{\mathcal{O}_\Xi}^{X \to X^{[n]}}(F^{\bullet})\,.$$

It is well known (see [Fog68]) that the Hilbert scheme $X^{[n]}$ of n points on X is a resolution of the singularities of $S^n X = X^n / \mathfrak{S}_n$ via the *Hilbert–Chow morphism*

$$\mu \colon X^{[n]} \to S^n X \quad , \quad \xi \mapsto \sum_{x \in \xi} \ell(\xi, x) \cdot x\,.$$

For every line bundle L on X the line bundle $L^{\boxtimes n} \in \mathrm{Pic}(X^n)$ descends to the line bundle $(L^{\boxtimes n})^{\mathfrak{S}_n}$ on $S^n X$. Thus for every $L \in \mathrm{Pic}(X)$ there is the *determinant line bundle* on $X^{[n]}$ given by

$$\mathcal{D}_L := \mu^*((L^{\boxtimes n})^{\mathfrak{S}_n})\,.$$

One goal in studying Hilbert schemes of points is to find formulas expressing the invariants of $X^{[n]}$ in terms of the invariants of the surface X. This includes the invariants of the induced sheaves defined above. There are already some results in this area. For example, in [Leh99] there is a formula for the Chern classes of $F^{[n]}$ in terms of those of F in the case that F is a line bundle. In [BNW07] the existence of universal formulas, i.e. formulas independent of the surface X, expressing the characteristic classes of any tautological sheaf in terms of the characteristic classes of F is shown and those formulas are computed in some cases. Furthermore Danila ([Dan01], [Dan07], [Dan00]) and Scala ([Sca09a], [Sca09b]) proved formulas for the cohomology of tautological sheaves, determinant line bundles, and some natural constructions (tensor, wedge, and symmetric products) of these. In particular in [Sca09a] there is the formula

$$\mathrm{H}^*(X^{[n]}, F^{[n]} \otimes \mathcal{D}_L) \cong \mathrm{H}^*(F \otimes L) \otimes S^{n-1}\, \mathrm{H}^*(L) \tag{1}$$

for the cohomology of a tautological sheaf twisted by a determinant line bundle. We will use and further develop Scala's approach of [Sca09a] and [Sca09b] which in turn uses the Bridgeland–King–Reid equivalence. Let G be a finite group acting on a smooth quasi-projective variety M. A *G-cluster* on M is a zero-dimensional closed G-invariant subscheme Z of M where $\Gamma(Z, \mathcal{O}_Z)$ equipped with the induced G-action is isomorphic to the natural representation $\mathbb{C}^G$. Bridgeland, King and Reid proved in [BKR01] that under some requirements the irreducible component $\mathrm{Hilb}^G(M)$, called the *Nakamura-G-Hilbert scheme*, of the fine moduli space of G-clusters on M which contains the points corresponding to free orbits is a crepant resolution of the quotient M/G. Furthermore, they showed that the G-equivariant Fourier–Mukai transform with kernel the structural sheaf of the universal family $\mathcal{Z}$ of G-clusters

$$\Phi := \Phi_{\mathcal{O}_{\mathcal{Z}}} \colon \mathrm{D}^b(\mathrm{Hilb}^G(M)) \to \mathrm{D}^b_G(M)$$

between the bounded derived category of $\mathrm{Hilb}^G(M)$ and the equivariant bounded derived category $\mathrm{D}^b_G(M) = \mathrm{D}^b(\mathrm{Coh}_G(M))$ of M is an equivalence of triangulated categories. Hence, Φ is called the *Bridgeland–King–Reid equivalence*. Haiman proved in [Hai01] that $X^{[n]}$ is isomorphic as a resolution of $S^n X$ to $\mathrm{Hilb}^{\mathfrak{S}_n}(X^n)$ with the *isospectral Hilbert scheme*

$$I^n X := \left(X^{[n]} \times_{S^n X} X^n\right)_{\mathrm{red}}$$

as the universal family of $\mathfrak{S}_n$-clusters. Furthermore, the conditions of the Bridgeland–King–Reid theorem are satisfied in this situation. Thus, there is the equivalence

$$\Phi := \Phi^{X^{[n]} \to X^n}_{\mathcal{O}_{I^n X}} \colon \mathrm{D}^b(X^{[n]}) \xrightarrow{\simeq} \mathrm{D}^b_{\mathfrak{S}_n}(X^n)\,.$$

In general for $\mathcal{E}^\bullet, \mathcal{F}^\bullet \in \mathrm{D}^b(\mathcal{A})$ objects in the derived category of any abelian category with enough injectives there is the identification $\mathrm{Ext}^k(\mathcal{E}^\bullet, \mathcal{F}^\bullet) \cong \mathrm{Hom}_{\mathrm{D}^b(\mathcal{A})}(\mathcal{E}^\bullet, \mathcal{F}^\bullet[k])$. Using this, we can compute extension groups on $X^{[n]}$ as $\mathfrak{S}_n$-invariant extension groups on X^n after applying the Bridgeland-King-Reid equivalence, i.e. for $\mathcal{E}^\bullet, \mathcal{F}^\bullet \in \mathrm{D}^b(X^{[n]})$ we have

$$\mathrm{Ext}^*_{X^{[n]}}(\mathcal{E}^\bullet, \mathcal{F}^\bullet) \cong \mathfrak{S}_n\, \mathrm{Ext}^*_{X^n}(\Phi(\mathcal{E}^\bullet), \Phi(\mathcal{F}^\bullet))\,.$$

Using the quotient morphism $\pi \colon X^n \to S^n X$ the right-hand side can be rewritten further as

$$\mathfrak{S}_n\, \mathrm{Ext}^*_{X^n}(\Phi(\mathcal{E}^\bullet), \Phi(\mathcal{F}^\bullet))) \cong \mathrm{H}^*(S^n X, [\pi_* R\mathcal{H}om_{X^n}(\Phi(\mathcal{E}^\bullet), \Phi(\mathcal{E}^\bullet)]^{\mathfrak{S}_n})\,.$$

Furthermore, there is a natural isomorphism $R\mu_*\mathcal{F}^\bullet \simeq [\pi_*\Phi(\mathcal{F}^\bullet)]^{\mathfrak{S}_n}$ (see [Sca09a]) which yields

$$\mathrm{H}^*(X^{[n]}, \mathcal{F}^\bullet) \cong \mathrm{H}^*(S^n X, [\pi_*\Phi(\mathcal{F}^\bullet)]^{\mathfrak{S}_n}) \cong \mathrm{H}^*(X^n, \Phi(\mathcal{F}^\bullet))^{\mathfrak{S}_n}. \tag{2}$$

So instead of computing the cohomology and extension groups of constructions of tautological sheaves on $X^{[n]}$ directly, the approach is to compute them for the image of these sheaves under the Bridgeland–King–Reid equivalence. In order to do this we need a good description of $\Phi(F^{[n]}) \in \mathrm{D}^b_{\mathfrak{S}_n}(X^n)$ for $F^{[n]}$ a tautological sheaf. This was provided by Scala in [Sca09a] and [Sca09b]. He showed that $\Phi(F^{[n]})$ is always concentrated in degree zero. This means that we can replace Φ by its non-derived version p_*q^* where p and q are the projections from $I^n X$ to X^n and $X^{[n]}$ respectively, i.e. we have $\Phi(F^{[n]}) \simeq p_*q^*(F^{[n]})$. Moreover, he gave for $p_*q^*(F^{[n]})$ a right resolution $C_F^\bullet$. This is a $\mathfrak{S}_n$-equivariant complex associated to F concentrated in non-negative degrees whose terms are *good* sheaves. For us a good $\mathfrak{S}_n$-equivariant sheaf on X^n is a sheaf which is constructed out of sheaves on the surface X in a not too complicated way. In particular, it should be possible to give a formula for its ($\mathfrak{S}_n$-invariant) cohomology in terms of the cohomology of sheaves on X. For example the degree zero term of the complex $C_F^\bullet$ is $C_F^0 = \bigoplus_{i=1}^n \mathrm{pr}_i^* F$. Note that if F is locally free C_F^0 is also. Its cohomology is by the Künneth formula given by

$$\mathrm{H}^*(X^n, C_F^0) = \left(\mathrm{H}^*(F) \otimes \mathrm{H}^*(\mathcal{O}_X)^{\otimes n-1}\right)^{\oplus n}.$$

The $\mathfrak{S}_n$-invariants of the cohomology can be computed as

$$\mathrm{H}^*(X^n, C_F^0)^{\mathfrak{S}_n} = \mathrm{H}^*(F) \otimes S^{n-1}\,\mathrm{H}^*(\mathcal{O}_X)$$

(For the proof of Scala's formula (1) in the case $L = \mathcal{O}_X$ it only remains to show that the invariants of C_F^p for $p \geq 1$ vanish). Let now $E_1, \ldots, E_k$ be locally free sheaves on X. The associated tautological sheaves $E_i^{[n]}$ on $X^{[n]}$ are again locally free and hence called *tautological bundles*. In [Sca09a] it is shown that again

$$\Phi(E_1^{[n]} \otimes \cdots \otimes E_k^{[n]}) \simeq p_*q^*(E_1^{[n]} \otimes \cdots \otimes E_k^{[n]}).$$

Furthermore, a description of $p_*q^*(\otimes_i E_i^{[n]})$ as the $E_\infty^{0,0}$ term of a certain spectral sequence is given. We will use a slightly different description of $p_*q^*(E_1^{[n]} \otimes \cdots \otimes E_k^{[n]})$ as a subsheaf of $K_0 := C_{E_1}^0 \otimes \cdots \otimes C_{E_k}^0$, namely as the image of $\otimes_{i=1}^k p_*q^*(E_i^{[n]})$ under the tensor product of the augmentation maps $\gamma_{E_i} \colon p_*q^*(E_i^{[n]}) \to C_{E_i}^0$. This allows us to proof the main result of chapter 3: We construct successively $\mathfrak{S}_n$-equivariant maps $\varphi_\ell \colon K_{\ell-1} \to T_\ell$ for $\ell = 1, \ldots, k$,

3

where $K_\ell := \ker \varphi_\ell$ and the T_ℓ are good sheaves given by

$$T_\ell = \bigoplus_{(M;i,j;a)} \left(S^{\ell-1}\Omega_X \otimes \bigotimes_{t \in M} E_t \right)_{ij} \otimes \bigotimes_{t \in [k] \setminus M} \mathrm{pr}^*_{a(t)} E_t.$$

The sum is taken over all tuples $(M; i, j; a)$ with $M \subset [k] := \{1, \ldots, k\}$, $|M| = \ell$, $i, j \in [n]$, $i \neq j$, and $a \colon [k] \setminus M \to [n]$. The functor $(\_)_{ij}$ is the composition $\iota_{ij*}p_{ij}^*$, where $\iota_{ij} \colon \Delta_{ij} \to X^n$ is the inclusion of the pairwise diagonal and $p_{ij} \colon \Delta_{ij} \to X$ is the restriction of the projection $\mathrm{pr}_i \colon X^n \to X$ respectively of $\mathrm{pr}_j \colon X^n \to X$. Then we show that $K_k = p_*q^*(\otimes_{i=1}^k E_i^{[n]})$. If the exact sequences

$$0 \to K_\ell \to K_{\ell-1} \to T_\ell$$

for $\ell = 1, \ldots, k$ were also exact with a zero on the right, this result would yield directly a description of the cohomology of $E_1^{[n]} \otimes \cdots \otimes E_k^{[n]}$ via long exact sequences and an explicit formula for its Euler characteristic. Since this is not the case, we have to enlarge the sequences to exact sequences with a zero on the right or at least do the same with the sequences

$$0 \to [\pi_* K_\ell]^{\mathfrak{S}_n} \to [\pi_* K_{\ell-1}]^{\mathfrak{S}_n} \to [\pi_* T_\ell]^{\mathfrak{S}_n}$$

of invariants on $S^n X$. The latter also yields the cohomological invariants since by (2) we have $\mathrm{H}^*(X^{[n]}, \otimes_i E_i^{[n]}) \cong \mathrm{H}^*(S^n X, [\pi_* K_k]^{\mathfrak{S}_n})$. By our approach we are able to give shorter and more unified proofs of the results of Danila and Scala. Thus, chapter 3 can also be seen as a survey of the present results on the cohomology of constructions of tautological bundles and determinant line bundles. The results are

- For a line bundle L on X and $k \leq n$ the formulas $\mu_*(\wedge^k L^{[n]}) \cong [\pi_* K_0 \otimes \mathfrak{a}_k]^{\mathfrak{S}_n \times \mathfrak{S}_k}$ where $\mathfrak{a}_k$ is the alternating presentation of $\mathfrak{S}_k$ and as a consequence the formula

$$\mathrm{H}^*(X^{[n]}, \wedge^k L^{[n]}) \cong \wedge^k \mathrm{H}^*(L) \otimes S^{n-k} \mathrm{H}^*(\mathcal{O}_X).$$

- For $k \leq n$ and L a line bundle on a projective surface X the formula

$$H^0(X^{[n]}, (L^{[n]})^{\otimes k}) \cong \mathrm{H}^0(L)^{\otimes k}$$

for the global sections of the tensor power of the associated tautological bundle.

- For E, F locally free sheaves on X a short exact sequence with a zero on the right and with $\mu_*(E^{[n]} \otimes F^{[n]})$ as the kernel and as a consequence a formula for the cohomology $\mathrm{H}^*(X^{[n]}, E^{[n]} \otimes F^{[n]})$.

Furthermore, we get the following new results

4

- For $E_1, \ldots, E_k$ locally free sheaves of arbitrary rank on a projective surface X and $k \leq n$ the formula
$$\mathrm{H}^0(X^{[n]}, E_1^{[n]} \otimes \cdots \otimes E_k^{[n]}) \cong \mathrm{H}^0(E_1) \otimes \cdots \otimes \mathrm{H}^0(E_k)$$

- For $E_1, \ldots, E_k$ locally free sheaves on X and arbitrary k long exact sequences
$$0 \to K_\ell \to K_{\ell-1} \to T_\ell \to T_\ell^1 \to \cdots \to T_\ell^{k-\ell} \to 0$$
on X^2 with good sheaves T_ℓ^i. This yields a description via long exact sequences of the cohomology and an explicit formula for the Euler characteristic of $E_1^{[2]} \otimes \cdots \otimes E_k^{[2]}$.

- Similar long exact sequences for arbitrary n over the open subset X_{**}^n consisting of points $(x_1, \ldots, x_n)$ where at most two x_i coincide.

- For E_1, E_2, E_3 locally free sheaves on X long exact sequences on $S^n X$ whose kernels converge to $[\pi_* K_3]^{\mathfrak{S}_n}$. This yields a description via long exact sequences of the cohomology and an explicit formula for the Euler characteristic of $E_1^{[n]} \otimes E_2^{[n]} \otimes E_3^{[n]}$.

Using that for every $\mathcal{F}^\bullet \in \mathrm{D}^b(X^{[n]})$ and L a line bundle on X we have
$$\Phi(\mathcal{F}^\bullet \otimes \mathcal{D}_L) \simeq \Phi(\mathcal{F}^\bullet) \otimes L^{\boxtimes n},$$

we can generalise the results from products of tautological bundles to products of tautological bundles twisted by a determinant line bundle by simply tensoring the exact sequences on X^n by $L^{\boxtimes n}$ respectively tensoring the exact sequences on $S^n X$ by $[\pi_* L^{\boxtimes n}]^{\mathfrak{S}_n}$. Also, we can generalise the results on the Euler characteristics form tautological bundles to arbitrary tautological objects. We will not state the results in this introduction since they are presented in a compact form in subsection 3.8.3.

In chapter 4 we compute extension groups between tautological bundles and more generally *twisted tautological objects*, i.e. tautological objects tensorised with determinant line bundles. Our main theorem is the existence of natural isomorphisms of graded vector spaces

$$\mathrm{Ext}^*((E^\bullet)^{[n]} \otimes \mathcal{D}_L, (F^\bullet)^{[n]} \otimes \mathcal{D}_M) \cong \begin{aligned} &\mathrm{Ext}^*(E^\bullet \otimes L, F^\bullet \otimes M) \otimes S^{n-1} \mathrm{Ext}^*(L, M) \oplus \\ &\mathrm{Ext}^*(E^\bullet \otimes L, M) \otimes \mathrm{Ext}^*(L, F^\bullet \otimes M) \otimes S^{n-2} \mathrm{Ext}^*(L, M) \end{aligned}$$

for objects $E^\bullet, F^\bullet \in \mathrm{D}^b(X)$ and line bundles $L, M \in \mathrm{Pic}(X)$. We also give a similar formula for $\mathrm{Ext}^*((E^\bullet)^{[n]} \otimes \mathcal{D}_L, \mathcal{D}_M)$. Since $\mathcal{D}_{\mathcal{O}_X} = \mathcal{O}_{X^{[n]}}$, by setting $L = M = \mathcal{O}_X$ the extension groups and the cohomology of the dual of non-twisted tautological objects occur as special cases. As an application of the formula for the extension groups we show for X a projective surface with a trivial canonical bundle that twisted tautological objects are never spherical or $\mathbb{P}^n$-objects in $\mathrm{D}^b(X^{[n]})$. The main step of the proof of the above formula is to compute

$[\pi_*R\,\mathcal{H}om_{X^n}(\Phi(E^{[n]}),\Phi(F^{[n]}))]^{\mathfrak{S}_n}$ in the case of tautological bundles, i.e. for E,F locally free sheaves on X. We show that

$$[\pi_*R\,\mathcal{H}om(\Phi(E^{[n]}),\Phi(F^{[n]})))]^{\mathfrak{S}_n} \simeq [\pi_*\,\mathcal{H}om(C_E^0,C_F^0)]^{\mathfrak{S}_n}. \tag{3}$$

In particular the $\mathfrak{S}_n$-invariants of the higher sheaf-Ext vanish. The isomorphism in degree zero is shown using the fact that the support of the terms C_E^p for $p \geq 0$ have codimension at least two and the normality of the variety X^n. For the vanishing of the higher derived sheaf-Homs we do computations in the spectral sequences associated to the bifunctor

$$\underline{\mathrm{Hom}}(\_,\_) := [\pi_*\,\mathcal{H}om_{X^n}(\_,\_)]^{\mathfrak{S}_n}$$

and the complexes $C_E^\bullet$ and $C_F^\bullet$. The formula (3) can easily be generalised to the case of tautological bundles twisted by determinant line bundles. We can generalise further to arbitrary objects $E^\bullet$, $F^\bullet$ by taking locally free resolutions on X and using some formal arguments for derived functors to get

$$\left[\pi_*R\,\mathcal{H}om(\Phi((E^\bullet)^{[n]}\otimes\mathcal{D}_L),\Phi((F^\bullet)^{[n]}\otimes\mathcal{D}_M))\right]^{\mathfrak{S}_n} \simeq \left[\pi_*R\,\mathcal{H}om(C_{E^\bullet}^0\otimes L^{\boxtimes n},C_{F_\bullet}^0\otimes M^{\boxtimes n})\right]^{\mathfrak{S}_n}.$$

Then we can compute the desired extension groups as the cohomology of the object on the right and get the main theorem. There is also a similar formula for the extension groups between two determinant line bundles and we can apply our arguments to generalise Scala's result to get a formula for the cohomology of twisted tautological objects. So now there are formulas for $\mathrm{Ext}^*_{X^{[n]}}(\mathcal{F}^\bullet,\mathcal{G}^\bullet)$ whenever both of $\mathcal{F}^\bullet$ and $\mathcal{G}^\bullet$ are either twisted tautological objects or determinant line bundles. We describe how to compute all the possible Yoneda products in terms of these formulas. Furthermore we give a interpretation of our results in terms of Čech cohomology on $X^{[n]}$ and compute the trace map and the cup product.

The author wants to thank his adviser Marc Nieper–Wißkirchen for his many valuable suggestions. He thanks Malte Wandel for some interseting discussions about the topic of this thesis. He also thanks Samuel Boissière and Marco Hien for their willingness to write expert's reports and to attend at the exam as members of the jury.

Contents

0.1 Introduction 1

1 Preliminaries   10
  1.1 General notations and conventions 10
  1.2 Combinatorical notations and preliminaries 12
     1.2.1 Symmetric groups 12
     1.2.2 Signs 12
     1.2.3 Multi-indices 13
     1.2.4 Partial diagonals 14
     1.2.5 Binomial coefficients 14
  1.3 Graded vector spaces and their Euler characteristics 15
     1.3.1 Graded vector spaces 15
     1.3.2 Grothendieck groups and the Euler characteristic 17
  1.4 Equivariant sheaves 18
     1.4.1 Basic definitions 18
     1.4.2 Inflation and Restriction 19
     1.4.3 Schemes with trivial G-action 20
     1.4.4 Equivariant geometric functors 20
     1.4.5 Derived equivariant categories 22
     1.4.6 Injective and locally free sheaves 22
     1.4.7 Derived equivariant functors 23
     1.4.8 Representations as G-sheaves 25
     1.4.9 Equivariant Grothendieck duality 25
  1.5 Preliminary lemmas 26
     1.5.1 Derived bifunctors 26
     1.5.2 Danila's lemma and corollaries 27
     1.5.3 Pull-back along regular embeddings 29
     1.5.4 Partial diagonals and the standard representation 33
     1.5.5 Normal varieties 38

2 Image of tautological sheaves under the Bridgeland-King-Reid equivalence 40

2.1 The Bridgeland–King–Reid equivalence 40

2.2 The Hilbert scheme of points on a surface 41

2.3 Tautological sheaves 43

2.4 The complex $C^\bullet$ 44

2.5 Polygraphs and the image of tautological sheaves under Φ 45

2.6 Description for $k \geq 2$ 48

2.7 Multitor spectral sequence 49

3 Cohomological invariants of twisted products of tautological sheaves 52

3.1 Description of $p_*q^*(E_1^{[n]} \otimes \cdots \otimes E_k^{[n]})$ 52

    3.1.1 Construction of the T_ℓ and φ_ℓ 52

    3.1.2 The open subset X_{**}^n 55

    3.1.3 Description of $p_*q^*(E_1^{[n]} \otimes \cdots \otimes E_k^{[n]})_{**}$ 56

    3.1.4 Description of $p_*q^*(E_1^{[n]} \otimes \cdots \otimes E_k^{[n]})$ 58

3.2 Invariants of K_0 and the T_ℓ 59

    3.2.1 Orbits and their isotropy groups on the sets of indices 59

    3.2.2 The sheaves of invariants and their cohomology 61

3.3 The map φ_1 on cohomology and the cup product 63

3.4 Cohomology in the highest and lowest degree 65

    3.4.1 Global sections for $n \geq k$ and X projective 65

    3.4.2 Cohomology in degree $2n$ 67

3.5 Wedge products in the case of line bundles 67

3.6 Tensor products of tautological bundles on $X^{[2]}$ and $X_{**}^{[n]}$ 69

    3.6.1 Long exact sequences on X^2 69

    3.6.2 The invariants on $S^2 X$ 77

    3.6.3 Cohomology on $X^{[2]}$ 78

    3.6.4 Long exact sequences on $X_{**}^{[n]}$ 80

3.7 Tensor products of tautological bundles on $X^{[n]}$ 83

    3.7.1 Restriction of local sections to closed subvarieties 83

    3.7.2 Double tensor products 85

    3.7.3 Triple tensor products 86

3.8 Generalisations 94

    3.8.1 Determinant line bundles 94

    3.8.2 Derived functors 95

    3.8.3 Generalised results 96

4 Extension groups of twisted tautological objects   **100**

4.1 The case of tautological bundles 101

    4.1.1 Computation of the $\underline{\mathrm{Hom}}$s 101

    4.1.2 Vanishing of the higher $\underline{\mathrm{Ext}}^i(\Phi(F^{[n]}), \mathcal{O}_{X^n})$ 102

    4.1.3 Vanishing of the higher $\underline{\mathrm{Ext}}^i(\Phi(E^{[n]}), \Phi(F^{[n]}))$ 103

4.2 Generalisations 108

    4.2.1 Determinant line bundles 108

    4.2.2 From tautological bundles to tautological objects 109

4.3 Global Ext-groups 110

4.4 Spherical and $\mathbb{P}^n$-objects 112

4.5 Products and interpretation of the results 114

    4.5.1 Yoneda products, the Künneth isomorphism and signs 114

    4.5.2 Yoneda products for twisted tautological objects 115

    4.5.3 Interpretation of the formulas 119

    4.5.4 The trace map and the cup product 122

References   **124**

Chapter 1

Preliminaries

1.1 General notations and conventions

(i) Given an abelian category $\mathcal{A}$ we will write $\cong$ for isomorphisms in $\mathcal{A}$ and $\simeq$ for isomorphisms in the derived category $\mathrm{D}(\mathcal{A})$. If an object $A \in \mathcal{A}$ shows up on one side of the sign $\simeq$ it is considered as the complex with A in degree 0 and vanishing terms elsewhere. If we want to emphasise that A is considered as the complex concentrated in degree zero, we will denote it by $A[0]$. More generally, $A[-i]$ is the complex concentrated in degree i with the only non-vanishing term being A. If we want to emphasise that an isomorphism is G-equivariant we will sometimes write $\cong_G$ respectively $\simeq_G$.

(ii) Let $C^\bullet \in \mathrm{D}^b(\mathcal{A})$ for any abelian category $\mathcal{A}$. We write $\mathcal{H}^i(C^\bullet)$ for the i-th cohomology of a complex, i.e. $\mathcal{H}^i(C^\bullet) = \ker(d^i)/\operatorname{im}(d^{i-1})$. If $\mathcal{A} = \operatorname{Coh}(X)$ is the category of coherent sheaves on a scheme X, we write $\mathrm{H}^i(X, C^\bullet)$ for the i-th (hyper-)cohomology of the complex of sheaves, i.e. $\mathrm{H}^i(X, \_) = R^i\Gamma(X, \_)$. We will often drop the X in the notation, i.e. we write $\mathrm{H}^i(\_) := \mathrm{H}^i(X, \_)$, especially when X is a fixed smooth quasi-projective surface.

(iii) For an exact functor $F\colon \mathcal{A} \to \mathcal{B}$ between abelian categories we write again F for the induced functor $\mathrm{D}(\mathcal{A}) \to \mathrm{D}(\mathcal{B})$ on the level of the derived categories. If we want to emphasise that the image of a complex $A^\bullet \in \mathrm{D}^b(\mathcal{A})$ under F is computed by applying $F\colon \mathcal{A} \to \mathcal{B}$ term-wise we write $(F(A))^\bullet$ instead of $F(A^\bullet)$. We also sometimes write $F(A^\bullet)$ in formulas for object in the derived categories when F is not exact although in this case F is not a functor between the derived categories. This means that we apply the functor $\operatorname{Kom}(F)$ to the complex $A^\bullet$, i.e. we apply F term-wise, and consider it again as an object in the derived category afterwards.

(iv) On a scheme X the sheaf-Hom functor is denoted by $\mathcal{H}om_{\mathcal{O}_X}$, $\mathcal{H}om_X$ or just $\mathcal{H}om$.

We write $(\_)^\vee = \mathcal{H}om(\_, \mathcal{O}_X)$ for the operation of taking the dual of a sheaf and $(\_)^{\mathrm{v}} = R\,\mathcal{H}om(\_, \mathcal{O}_X)$ for the derived dual.

(v) All varieties are reduced and irreducible.

(vi) We will often write the symbol "PF" above an isomorphism symbol to indicate that the isomorphism is given by the projection formula. Also we use "lf" when the given isomorphism is because of some sheaf being locally free and thus a tensor product or sheaf-Hom functor needs not to be derived or commutes with taking cohomology.

(vii) For a finite vector space V we will write $v := \dim V$.

(viii) An empty tensor, wedge or symmetric product of sheaves on X is the sheaf $\mathcal{O}_X$.

(ix) Given a product of schemes $X = \prod_{i \in I} X_i$ we will denote by pr_i, pr_{X_i}, p_i, or p_{X_i} the projection $X \to X_i$.

(x) In formulas with enumerations putting the sign $\hat{}$ over an element means that this element is omitted. For example $\{1, \ldots, \hat{3}, \ldots, 5\}$ denotes the set $\{1, 2, 4, 5\}$.

(xi) For a local section s of a sheaf $\mathcal{F}$ we will often write $s \in \mathcal{F}$.

(xii) For a direct sum $V = \oplus_{i \in I} V_i$ of vector spaces or sheaves we will write interchangeably V_i and $V(i)$ for the summands. For an element respectively local section $s \in V$ we will write $s(i)$ or s_i for its component in $V(i)$. We denote the components of a morphism $\psi \colon Z \to V$ by $\psi(i) \colon Z \to V(i)$. Let $W = \oplus_{j \in J} W_j$ be an other direct sum and $\varphi \colon V \to W$ a morphism. We will denote the components of φ by $\varphi(i, j) \colon V(i) \to W(j)$.

(xiii) Let $\iota \colon Z \to X$ be a closed embedding of schemes and let $F \in \mathrm{QCoh}(X)$ be a quasi-coherent sheaf on X. The symbol $F_{|Z}$ will sometimes denote the sheaf $\iota^* F \in \mathrm{QCoh}(Z)$ and at other times the sheaf $\iota_* \iota^* F \in \mathrm{QCoh}(X)$. The restriction morphism

$$F \to F_{|Z} = \iota_* \iota^* F$$

is the unit of the adjunction (ι^*, ι_*). The image of a section $s \in F$ under this morphism is denoted by $s_{|Z}$.

1.2 Combinatorical notations and preliminaries

1.2.1 Symmetric groups

For any finite set M the symmetric group $\mathfrak{S}_M$ is the group of bijections of M. Note that we have $\mathfrak{S}_\emptyset \cong 1$. For two positive integers $n < m$ we use the notation

$$[n] = \{1, 2, \ldots, n\} \quad , \quad [n, m] = \{n, n+1, \ldots, m\}.$$

If $n > m$ we set $[n, m] := \emptyset$. We interpret the symmetric group $\mathfrak{S}_n$ as the group acting on $[n]$, i.e. $\mathfrak{S}_n = \mathfrak{S}_{[n]}$. For any subset $I \subset [n]$ we denote by $\bar{I} = [n] \setminus I$ its complement in $[n]$. For better readability we sometimes write $\overline{\mathfrak{S}_I}$ instead of $\mathfrak{S}_{\bar{I}}$.

1.2.2 Signs

Let M be a finite set. There is the group homomorphism sgn: $\mathfrak{S}_M \to \{-1, +1\}$ which is given after choosing a total order $<$ on M by

$$\operatorname{sgn} \sigma = (-1)^{\#\{(i,j) \in M \times M \mid i < j, \sigma(i) > \sigma(j)\}}$$

for $\sigma \in \mathfrak{S}_M$. For two finite totally ordered sets M, L of the same cardinality we define $u_{M \to L}$ as the unique strictly increasing map. Let now N be totally ordered, $m \in M \subset N$ and $\sigma \in \mathfrak{S}_N$. We define the signs

$$\varepsilon_{\sigma,M} := \operatorname{sgn}(u_{\sigma(M) \to M} \circ \sigma_{|M}) = (-1)^{\#\{(i,j) \in M \times M \mid i < j, \sigma(i) > \sigma(j)\}}$$

and $\varepsilon_{m,M} := (-1)^{\#\{j \in M \mid j < m\}}$.

Lemma 1.2.1. *(i) Let L, M be two finite totally ordered sets. For $\sigma \in \mathfrak{S}_M$ we consider $\tilde{\sigma} := u_{M \to L} \circ \sigma \circ u_{L \to M} \in \mathfrak{S}_L$. Then $\operatorname{sgn} \sigma = \operatorname{sgn} \tilde{\sigma}$.*

(ii) Let N be a finite set with a total order and $M \subset N$. Then

$$\varepsilon_{\mu,\sigma(M)} \cdot \varepsilon_{\sigma,M} = \varepsilon_{\mu \circ \sigma, M}$$

for all $\sigma, \mu \in \mathfrak{S}_N$.

(iii) Let $M \subset N$ be as above and $m \in M$. Then

$$\varepsilon_{\sigma^{-1}(m),\sigma^{-1}(M)} \cdot \varepsilon_{\sigma,\sigma^{-1}(M)} = \varepsilon_{m,M} \cdot \varepsilon_{\sigma,\sigma^{-1}(M \setminus \{m\})}$$

holds for every $\sigma \in \mathfrak{S}_N$.

Proof. The map $u_{M \to L} \times u_{M \to L}$ gives a bijection

$$\{(i,j) \in M \times M \mid i < j,\, \sigma(i) > \sigma(j)\} \cong \{(i,j) \in L \times L \mid i < j,\, \tilde{\sigma}(i) > \tilde{\sigma}(j)\},$$

which proves (i). For (ii) we have

$$
\begin{aligned}
\varepsilon_{\mu,\sigma(M)} \cdot \varepsilon_{\sigma,M} &= \operatorname{sgn}(u_{\mu(\sigma(M)) \to \sigma(M)} \circ \mu_{|\sigma(M)}) \operatorname{sgn}(u_{\sigma(M) \to M} \circ \sigma_{|M}) \\
&\overset{(i)}{=} \operatorname{sgn}(u_{\sigma(M) \to M} u_{\mu(\sigma(M)) \to \sigma(M)} \mu_{|\sigma(M)} u_{M \to \sigma(M)}) \operatorname{sgn}(u_{\sigma(M) \to M} \circ \sigma_{|M}) \\
&= \operatorname{sgn}(u_{\sigma(M) \to M} u_{\mu(\sigma(M)) \to \sigma(M)} \mu_{|\sigma(M)} u_{M \to \sigma(M)} u_{\sigma(M) \to M} \sigma_{|M}) \\
&= \operatorname{sgn}\left(u_{\mu(\sigma(M)) \to M}(\mu \circ \sigma)_{|M}\right) \\
&= \varepsilon_{\mu \circ \sigma, M}.
\end{aligned}
$$

For $m \in M \subset N$ and $\sigma \in \mathfrak{S}_N$ as in (iii) we have

$$
\begin{aligned}
&|\{j \in \sigma^{-1}(M) \mid j < \sigma^{-1}(m)\}| - |\{j \in M \mid j < m\}| \\
={}&|\{j \in \sigma^{-1}(M) \mid j < \sigma^{-1}(m), \sigma(j) > m\}| - |\{j \in \sigma^{-1}(M) \mid j > \sigma^{-1}(m), \sigma(j) < m\}|.
\end{aligned}
$$

This yields

$$\frac{\varepsilon_{\sigma^{-1}(m),\sigma^{-1}(M)}}{\varepsilon_{m,M}} = \frac{\varepsilon_{\sigma,\sigma^{-1}(M)}}{\varepsilon_{\sigma,\sigma^{-1}(M \setminus \{m\})}} = \frac{\varepsilon_{\sigma,\sigma^{-1}(M \setminus \{m\})}}{\varepsilon_{\sigma,\sigma^{-1}(M)}}.$$

$\square$

1.2.3 Multi-indices

We will write multi-indices mostly in the form of maps, i.e. for two numbers $n, k \in \mathbb{N}$ we denote multi-indices with k values between 1 and n rather as elements of $\operatorname{Map}([k],[n])$ than as elements of $[n]^k$. But sometimes we will switch between the notations and write a multi-index $a \colon [k] \to [n]$ in the form $a = (a(1), \ldots, a(k))$ or $a = (a_1, \ldots, a_k)$. For two maps $a \colon M \to K$ and $b \colon N \to K$ with disjoint domains we write $a \uplus b \colon M \coprod N \to K$ for the induced map on the union. If $N = \{i\}$ consists of only one element we will also write $a \uplus b = (a, i \mapsto b(i))$. For $x \in K$ we write $\underline{x} \colon M \to K$ for the map which is constantly x. For a multi-index $a \colon M \to \{i < j\}$ with a totally ordered codomain consisting of two elements we introduce the sign

$$\varepsilon_a := (-1)^{\# a^{-1}(\{j\})}.$$

For the preimage sets of one element i in the codomain of a we will often write for short $a^{-1}(i)$ instead of $a^{-1}(\{i\})$.

1.2.4 Partial diagonals

Let X be a variety and $n \in \mathbb{N}$. We define for $I \subset [n]$ with $|I| \geq 2$ the *I-th partial diagonal* as the reduced closed subvariety given by

$$\Delta_I = \{(x_1, \ldots, x_n) \in X^n \mid x_i = x_j \, \forall \, i, j \in I\} \, .$$

We denote by $\mathrm{pr}_i \colon X^n \to X$ the projection on the i-th factor and by $p_I \colon \Delta_I \to X$ the projection induced by pr_i for any $i \in I$. We denote the inclusion of the partial diagonals into the product by $\iota_I \colon \Delta_I \to X^n$. For a coherent sheaf F on X we set

$$F_I := \iota_{I*} p_I^* F \, .$$

The functor $(\_)_I \colon \mathrm{Coh}(X) \to \mathrm{Coh}(X^n)$ is an exact functor since it is a pull-back along a projection followed by a push-forward along a closed embedding. We will sometimes drop the brackets $\{\_\}$ in the notations, e.g we will write

$$\Delta_{12} = \Delta_{1,2} = \Delta_{\{1,2\}} \quad , \quad F_{12} = F_{1,2} = F_{\{1,2\}} \, .$$

Moreover, we denote the vanishing ideal sheaf of Δ_I in X^n by $\mathcal{I}_I$. If X is non-singular we denote the normal bundle of the partial diagonal by $N_I := N_{\Delta_I} = (\iota_I^* \mathcal{I}_I)^\vee = (\mathcal{I}_I / \mathcal{I}_I^2)^\vee$. For $|I| \in \{0,1\}$ we set $\Delta_I = X^n$. The *big diagonal* in X^n is defined as $\mathbb{D} := \cup_{1 \leq i < j \leq n} \Delta_{ij}$. The *small diagonal* is $\Delta = \Delta_{[n]}$.

1.2.5 Binomial coefficients

For $n, k \in \mathbb{Z}$ with $k \geq 0$ the binomial coefficient is given by

$$\binom{n}{k} = \frac{1}{k!} \cdot n(n-1) \cdots (n-k+1) \, .$$

In particular for $0 \leq n \leq k$ it is zero.

Lemma 1.2.2. *For all $k, \ell \in \mathbb{N}$ with $k \geq \ell$ we have*

$$\sum_{i=0}^{k-\ell} (-1)^i 2^{k-\ell-i} \binom{k}{\ell+i} \binom{\ell+i-1}{\ell-1} = \sum_{j=\ell}^{k} \binom{k}{j} \, .$$

Proof. For $k = \ell$ both sides of the equation equal 1. Inserting $k+1$ for k we get by induction

$$\sum_{i=0}^{k-\ell+1} (-1)^i 2^{k-\ell-i+1} \binom{k+1}{\ell+i} \binom{\ell+i-1}{\ell-1}$$

14

$$= \sum_{i=0}^{k-\ell+1} (-1)^i 2^{k-\ell-i+1} \left(\binom{k}{\ell+i} + \binom{k}{\ell+i-1} \right) \binom{\ell+i-1}{\ell-1}$$

$$= 2\sum_{i=0}^{k-\ell} (-1)^i 2^{k-\ell-i} \binom{k}{\ell+i} \binom{\ell+i-1}{\ell-1}$$

$$+ \sum_{i=0}^{k-\ell+1} (-1)^i 2^{k-\ell-i+1} \binom{k}{\ell+i-1} \binom{\ell+i-1}{\ell-1}$$

$$= 2\sum_{j=\ell}^{k} \binom{k}{j} + 2^{k-\ell+1} \binom{k}{\ell-1} \sum_{i=0}^{k-\ell+1} (-2)^{-i} \binom{k-\ell+1}{i}$$

$$= 2\sum_{j=\ell}^{k} \binom{k}{j} + 2^{k-\ell+1} \binom{k}{\ell-1} (1 - \frac{1}{2})^{k-\ell+1}$$

$$= 2\sum_{j=\ell}^{k} \binom{k}{j} + \binom{k}{\ell-1}$$

$$= \sum_{j=\ell}^{k+1} \left(\binom{k}{j} + \binom{k}{j-1} \right)$$

$$= \sum_{j=\ell}^{k+1} \binom{k+1}{j}.$$

$\square$

Lemma 1.2.3. *For $\chi, m \in \mathbb{Z}$ with $m \geq 0$ there is the equation*

$$(-1)^m \binom{-\chi}{m} = \binom{\chi+m-1}{m}.$$

Proof. Indeed we have

$$m! \cdot \binom{-\chi}{m} = (-\chi)\cdots(-\chi-m+1) = (-1)^m \chi \cdots (\chi+m-1) = (-1)^m m! \binom{\chi+m-1}{m}.$$

$\square$

1.3 Graded vector spaces and their Euler characteristics

In this section all vector spaces will be vector spaces over a field k of characteristic zero.

1.3.1 Graded vector spaces

A *graded vector space* V^* is a vector space together with a decomposition $V^* = \oplus_{i \in \mathbb{Z}} V^i$ where only finitely many V^i are non-zero. A *super vector space* $V^{\pm}$ is a vector space together with

15

a decomposition $V^\pm = V^+ \oplus V^-$. There is a functor $G\colon \mathrm{GVS} \to \mathrm{SVS}$ from the category of graded vector spaces to the category of super vector spaces given by $V^* \mapsto V^\pm$ with

$$V^+ = \bigoplus_{i \text{ even}} V^i \quad, \quad V^- = \bigoplus_{i \text{ odd}} V^i.$$

It has a right quasi-inverse F given by $V^\pm \mapsto V^*$ with $V^0 = V^+$, $V^1 = V^-$, and $V^i = 0$ for $i \notin \{0, 1\}$. There is also a functor $\mathrm{GVS} \to \mathrm{D}^b(\mathrm{Vec})$ to the bounded derived category of vector spaces given by $V^* \mapsto V^\bullet$, where $V^\bullet$ is the complex with trivial differentials and whose term in degree i is V^i. Since Vec is a semi-simple category, this is an equivalence with quasi-inverse given by $C^\bullet \mapsto C^*$ with $C^i = \mathcal{H}^i(C^\bullet)$ (see [GM96, II. 2.3]). Thus, we can define the tensor product of graded vector spaces as the tensor product of the associated complexes. This means that for two graded vector spaces V^* and W^* the tensor product is given by $(V \otimes W)^i = \oplus_{p+q=i} V^p \otimes W^q$. For two super vector spaces $V^\pm$ and $W^\pm$ we define their tensor product by $V^\pm \otimes W^\pm := G(F(V^\pm) \otimes F(W^\pm))$. More concretely, we have $(V \otimes W)^+ = (V^+ \otimes W^+) \oplus (V^- \otimes W^-)$ and $(V \otimes W)^- = (V^+ \otimes W^-) \oplus (V^- \otimes W^+)$. As one can check, the tensor power $(C^\bullet)^{\otimes m}$ of a complex $C^\bullet \in \mathrm{D}^b(\mathrm{Vec})$ becomes a $\mathfrak{S}_m$-equivariant complex with the action given by

$$\sigma \cdot (u_1 \otimes \cdots \otimes u_m) := \varepsilon_{\sigma, p_1, \ldots, p_m} (u_{\sigma^{-1}(1)} \otimes \cdots \otimes u_{\sigma^{-1}(m)}).$$

Here the p_i are the degrees of the u_i, i.e. $u_i \in C^{p_i}$ and the sign $\varepsilon_{\sigma, p_1, \ldots, p_m}$ is defined by setting $\varepsilon_{\tau, p_1, \ldots, p_m} = (-1)^{p_i \cdot p_{i+1}}$ for the transposition $\tau = (i, i+1)$ and requiring it to be a homomorphism in σ. We call $\varepsilon_{\sigma, p_1, \ldots, p_m}$ the *cohomological sign*. Consequently, we define the symmetric product $S^m C^\bullet$ respectively $S_m C^\bullet$ degree-wise as the coinvariants respectively the invariants of $(C^\bullet)^{\otimes m}$ under this action. Since k is of characteristic zero, there is the natural isomorphism $S^m C^\bullet \to S_m C^\bullet$ given by

$$u_1 \cdots u_m \mapsto \frac{1}{m!} \sum_{\sigma \in \mathfrak{S}_m} \varepsilon_{\sigma, p_1, \ldots, p_m} (u_{\sigma^{-1}(1)} \otimes \cdots \otimes u_{\sigma^{-1}(m)}).$$

Again, we define the symmetric product of graded vector spaces as the symmetric product of the associated complex and for a super vector space $V^\pm$ we set $S^m V^\pm := G(S^m(F(V^\pm)))$. Similarly, we define $\wedge^m C^\bullet$ and $\wedge_m C^\bullet$ as the coinvariants respectively invariants by the $\mathfrak{S}_m$-action on $(C^\bullet)^{\otimes m}$ from above twisted by the alternating representation, i.e. by the action

$$\sigma \cdot (u_1 \otimes \cdots \otimes u_m) := \mathrm{sgn}(\sigma) \varepsilon_{\sigma, p_1, \ldots, p_m} (u_{\sigma^{-1}(1)} \otimes \cdots \otimes u_{\sigma^{-1}(m)}).$$

Again, this leads to the definition of the wedge product of graded and super vector spaces.

16

1.3.2 Grothendieck groups and the Euler characteristic

For an abelian category $\mathcal{A}$ we define the *Grothendieck group* $\mathrm{K}(\mathcal{A})$ as the abelian group

$$\mathbb{Z} \cdot \{ \text{ isomorphism classes } [A] \text{ of objects } A \in \mathcal{A} \}/H$$

with H being the subgroup generated by elements of the form $[A] - [A'] - [A'']$ for exact sequences

$$0 \to A' \to A \to A'' \to 0\,.$$

We define a map $[\_] \colon \mathrm{D}^b(\mathcal{A}) \to \mathrm{K}(\mathcal{A})$ for $A^\bullet \in \mathrm{D}^b(\mathcal{A})$ by

$$[A^\bullet] = \sum_{i \in \mathbb{Z}} (-1)^i [A^i] = \sum_{i \in \mathbb{Z}} (-1)^i \mathcal{H}^i(A^\bullet)\,.$$

For a scheme X we set $\mathrm{K}(X) := \mathrm{K}(\mathrm{Coh}(X))$ and $\mathrm{K}^0(X) := \mathrm{K}(\mathrm{LF}(X))$, where $\mathrm{LF}(X)$ denotes the category of locally free sheaves on X. Whenever there are enough locally free sheaves on X, both groups are isomorphic. The group $\mathrm{K}(X)$ is a $\mathrm{K}^0(X)$-module with the multiplication for $E \in \mathrm{LF}(X)$ and $F \in \mathrm{Coh}(X)$ given by $[E] \cdot [F] := [E \otimes F]$. For a proper morphism $\mu \colon X \to Y$ we define the push-forward $\mu_!$ for $F \in \mathrm{Coh}(X)$ by

$$\mu_![F] := \sum_{i \in \mathbb{Z}} (-1)^i R^i \mu_*(F)\,.$$

Equivalently, we can consider F as a complex in $\mathrm{D}^b(X)$ concentrated in degree zero and apply the composition $[\_] \circ R\mu_*$. For a bounded complex $C^\bullet \in \mathrm{D}^b(\mathrm{Vec}_f)$ of finite vector spaces its *Euler characteristic* is given by

$$\chi(C^\bullet) := \chi(V^*) := \sum_{i \in \mathbb{Z}} (-1)^i c^i = \sum_{i \in \mathbb{Z}} (-1)^i \dim \mathcal{H}^i(C^\bullet) = \sum_{i \in \mathbb{Z}} (-1)^i v^i = v^+ - v^-\,.$$

Here we use the convention of 1.1 (vii) and denote the graded vector space associated to $C^\bullet$ by V^* and the associated super vector space by $V^\pm$. The third equality makes sure that the Euler characteristic is invariant under isomorphisms in $\mathrm{D}^b(\mathrm{Vec}_f)$. The Euler characteristic of a finite graded vector space is defined as the Euler characteristic of the associated complex. Let X be a complete k-scheme. For a sheaf $F \in \mathrm{Coh}(X)$ or more generally an object $F \in \mathrm{D}^b(X)$ we set $\chi(F) := \chi(\mathrm{H}^*(X, F))$. The Euler characteristic of F is an invariant of $[F] \in \mathrm{K}(X)$.

Lemma 1.3.1. *Let V^* and W^* be graded vector spaces and $m \in \mathbb{N}$. Then there are the formulas*

$$\chi(V^* \otimes W^*) := \chi(V^*) \cdot \chi(W^*) \quad , \quad \chi(S^m V^*) = \binom{\chi(V^*) + m - 1}{m}\,.$$

Proof. Using the description of the tensor product $V^\pm \otimes W^\pm$ in the previous subsection, the first formula follows from the equality

$$(v^+w^+ - v^-w^-) - (v^+w^- + v^-w^+) = (v^+ - v^-)(w^+ - w^-).$$

For the second formula we first consider the case that $\chi := \chi(V^*) \geq 0$, i.e. $v^+ \geq v^-$. We can construct a short exact sequence

$$0 \to k^\chi \to V^- \xrightarrow{d} V^- \to 0.$$

This gives an isomorphism $k^\chi[0] \simeq V^\bullet$, where $V^\bullet$ is the complex concentrated in degree 0 and 1 given by $d \colon V^+ \to V^-$. Since the symmetric product of complexes preserves isomorphisms in $D^b(\mathrm{Vec}_f)$, we get

$$\chi(S^m V^*) = \chi(S^m V^\bullet) = \chi(S^m(k^\chi[0])) = \dim S^m k^\chi = \binom{\chi + m - 1}{m}.$$

In the case that $\chi < 0$ we can construct a short exact sequence

$$0 \to V^+ \xrightarrow{d} V^- \to k^{-\chi} \to 0$$

which is the same as an isomorphism $V^\bullet \simeq k^{-\chi}[-1]$. The complex $S^m(k^{-\chi}[-1])$ equals $(\wedge^m k^{-\chi})[-m]$. Thus,

$$\chi(S^m V^*) = \chi(S^m V^\bullet) = \chi(S^m(k^{-\chi}[-1])) = (-1)^m \dim(\wedge^m k^{-\chi}) = (-1)^m \binom{-\chi}{m}$$

$$\overset{1.2.3}{=} \binom{\chi + m - 1}{m}.$$

$\square$

1.4 Equivariant sheaves

1.4.1 Basic definitions

Let G be a finite group acting on a scheme X. All group actions will be left actions. Let $F \in \mathrm{QCoh}(X)$ be a quasi-coherent sheaf. A *G-linearization* on F is a family of $\mathcal{O}_X$-linear isomorphisms $(\lambda_g \colon F \to g^* F)_{g \in G}$ with the properties $\lambda_e = \mathrm{id}$ and

$$\lambda_{hg} = g^* \lambda_h \circ \lambda_g \colon F \to g^* h^* F \cong (hg)^* F.$$

A quasi-coherent sheaf $F \in \mathrm{QCoh}(X)$ together with a G-linearization is called a G-*equivariant sheaf* or just a G-*sheaf*. For two G-sheaves (E, λ), (F, μ) the group G acts on $\mathrm{Hom}_{\mathcal{O}_X}(E, F)$ by

$$g \cdot \varphi := \mu_{g^{-1}}^{-1} \circ ((g^{-1})^* \varphi) \circ \lambda_{g^{-1}} \,.$$

A *morphism of G-sheaves* is an $\mathcal{O}_X$-linear morphism of the underlying ordinary sheaves which is invariant under this action, i.e.

$$G \, \mathrm{Hom}_{\mathcal{O}_X}(E, F) := (\mathrm{Hom}_{\mathcal{O}_X}(E, F))^G = \{\varphi \colon E \to F \mid \mu_g \circ \varphi = (g^* \varphi) \circ \lambda_g \; \forall g \in G\} \,.$$

This gives the abelian category $\mathrm{QCoh}_G(X)$ of G-equivariant sheaves on X. The full abelian subcategory of G-sheaves whose underlying sheaves are coherent is denoted by $\mathrm{Coh}_G(X)$.

1.4.2 Inflation and Restriction

For every subgroup $H \subset G$ we can restrict the G-linearization of a G-sheaf F to a H-linearization, which yields the *restriction functor* $\mathrm{Res}_G^H \colon \mathrm{QCoh}_G(X) \to \mathrm{QCoh}_H(X)$. In the special case $H = 1$ we call the functor

$$\mathrm{For} := \mathrm{Res}_G^1 \colon \mathrm{QCoh}_G(X) \to \mathrm{QCoh}(X)$$

also the *forgetful functor*. We choose a system of representatives of $H \setminus G$. Then there is the *inflation functor*

$$\mathrm{Inf}_H^G \colon \mathrm{QCoh}_H(X) \to \mathrm{QCoh}_G(X) \quad , \quad \mathrm{Inf}_H^G(E, \lambda) := \bigoplus_{[g] \in H \setminus G} g^* E$$

depending only by canonical isomorphism on the chosen representatives. The G-linearization μ of $\mathrm{Inf}_H^G(E)$ is given as follows. For every $\sigma \in G$ and g one of the chosen representatives of the classes in $H \setminus G$ there exist uniquely $h \in H$ and $\hat{g}$ one of the chosen representatives such that $h \cdot \hat{g} = g \cdot \sigma$. Now for a local section $s = (s_g \in g^* E)_{[g] \in H \setminus G}$ we define $\mu_\sigma(s)_g \in \sigma^* g^* E$ as the image of $s_{\hat{g}}$ under the isomorphism

$$\hat{g}^* E \overset{\hat{g}^* \lambda_h}{\to} \hat{g}^* h^* E \cong (h\hat{g})^* E = (g\sigma)^* E \cong \sigma^* g^* E \,.$$

The inflation functor is left adjoint to the restriction functor. The adjoint pair

$$\mathrm{Inf}_H^G \colon \mathrm{QCoh}_H(X) \rightleftarrows \mathrm{QCoh}_G(X) \colon \mathrm{Res}_G^H$$

clearly restricts to an adjoint pair of the full subcategories of equivariant coherent sheaves. Both the inflation and the restriction functor are exact. Let G act transitively on a set

19

I. Let $\mathcal{M}$ be a G-sheaf on X where the underlying ordinary sheaf admits a decomposition $\mathcal{M} = \oplus_{i \in I} \mathcal{M}_i$ such that for any $i \in I$ and $g \in G$ the linearization λ restricted to $\mathcal{M}_i$ is an isomorphism $\lambda_g \colon \mathcal{M}_i \overset{\cong}{\to} g^* \mathcal{M}_{g(i)}$. Then the G-linearization of $\mathcal{M}$ restricts to a $\mathrm{Stab}_G(i)$-linearization of $\mathcal{M}_i$.

Lemma 1.4.1. *Under these assumptions for every $i \in I$ we have $\mathcal{M} \cong_G \mathrm{Inf}^G_{\mathrm{Stab}(i)} \mathcal{M}_i$.*

Proof. We only have to show that $\mathcal{M}$ fulfils the adjointness property of $\mathrm{Inf}^G_{\mathrm{Stab}(i)}(\mathcal{M}_i)$. Let $(\mathcal{F}, \mu)$ be a G-sheaf and $\varphi \colon \mathcal{M} \to \mathcal{F}$ a G-equivariant morphism. For $j \in I$ let $\varphi_j \colon \mathcal{M}_j \to \mathcal{F}$ be the component of φ. Choose a $g \in G$ with $g(j) = i$. Then by the G-equivariance

$$\varphi_j = \mu_g^{-1} \circ (g^* \varphi_i) \circ (\lambda_{g | \mathcal{M}_j}).$$

Thus φ is determined by the $\mathrm{Stab}(i)$-equivariant morphism φ_i. On the other hand every given $\mathrm{Stab}(i)$-equivariant morphism $\varphi_i \colon \mathcal{M}_i \to \mathcal{F}$ gives rise to a G-equivariant $\varphi \colon \mathcal{M} \to \mathcal{F}$ by taking the above equation as the definition of the other components. $\qquad\square$

Remark 1.4.2. Conversely the inflated sheaf as defined above can be written in the form of $\mathcal{M}$ from above by setting $I = H \setminus G$ with $\sigma \in G$ acting on I by $\sigma(Hg) = Hg\sigma^{-1}$.

1.4.3 Schemes with trivial G-action

If G acts trivially on X, a G-linearization of F is just a G-action on F, i.e. a family of G-actions on $F(U)$ for each open subset $U \subset X$, which are compatible with the restrictions. Thus, in this case there is also the *functor of taking invariants* $[\_]^G \colon \mathrm{QCoh}_G(X) \to \mathrm{QCoh}(X)$ given by $F^G(U) := [F(U)]^G$. If the scheme is defined over a field of characteristic zero the functor of taking invariants is exact because of the existence of the Reynolds operator. Considering an ordinary sheaf on X as a sheaf with the trivial G-action also yields an exact functor $\mathrm{triv} \colon \mathrm{QCoh}(X) \to \mathrm{QCoh}_G(X)$. Together these functors form the adjoint pair

$$\mathrm{triv} \colon \mathrm{QCoh}(X) \rightleftharpoons \mathrm{QCoh}_G(X) \colon [\_]^G$$

Again both functors restrict to functors between the coherent categories. We will write interchangeably $[\_]^G$ and $(\_)^G$ or leave the brackets away.

1.4.4 Equivariant geometric functors

Let G act also on another scheme Y and $f \colon X \to Y$ be an G-equivariant morphism. Then there are the *equivariant geometric functors*

$$\otimes, \mathcal{H}om \colon \mathrm{QCoh}_G(X) \times \mathrm{QCoh}_G(X) \to \mathrm{QCoh}_G(X),$$

$f_*\colon \mathrm{QCoh}_G(X) \to \mathrm{QCoh}_G(Y)$ and $f^*\colon \mathrm{QCoh}_G(Y) \to \mathrm{QCoh}_G(X)$ which are defined in a canonical way such that they are compatible with the corresponding non-equivariant functors via the forgetful functor. More concretely, since f commutes with the action of every $\sigma \in G$ for $D, E \in \mathrm{QCoh}(X)$ and $F \in \mathrm{QCoh}(Y)$ there are natural isomorphisms $f^*\sigma^*F \cong \sigma^*f^*F$, $f_*\sigma^*E \cong \sigma^*f_*E$ (flat base change) and $\sigma^*D \otimes \sigma^*E \cong \sigma^*(D \otimes E)$. This allows us to define the G-linearization of the pull-back, the push-forward and the tensor product as the pull-back, push-forward and tensor product of the G-linearizations of the original sheaves. Since the morphism σ is an automorphism the natural morphism $\sigma^*\mathcal{H}om(D, E) \to \mathcal{H}om(\sigma^*D, \sigma^*E)$ is also an isomorphism. Using this identification, the linearization α of $\mathcal{H}om(D, E)$ is given by $\alpha_\sigma(\varphi) = (\mu_{\sigma|U}) \circ \varphi \circ (\lambda_{\sigma|U}^{-1})$ where $\varphi \in \mathcal{H}om(D, E)(U)$ for an open subset $U \subset X$ and λ and μ are the linearizations of D and E. Because of the compatibility with the forgetful functor, all these functors restrict to functors between the full subcategories of coherent sheaves if and only if the corresponding non-equivariant geometric functors do. Every G-linearization induces a G-action on the global sections of X. Hence, there is also the functor

$$\Gamma(X, \_)\colon \mathrm{QCoh}(X) \to \mathrm{Mod}(\mathbb{Z}[G])$$

mapping to the category of abelian groups equipped with an additive G-action. If X is defined over a field k we always assume the structural morphism $X \to \mathrm{Spec}\, k$ to be G-invariant, i.e G acts k-linear on X. Thus, in this case the global sections functor factorises over the category of $k[G]$-modules. We have the formula

$$G\,\mathrm{Hom}(D, E) = [\Gamma(X, \mathcal{H}om(D, E))]^G\,.$$

Lemma 1.4.3. *Let $H \leq G$ be a subgroup, $\mathcal{E}$ a G-sheaf and $\mathcal{F}$ a H-sheaf on X. We set $\mathrm{Inf} = \mathrm{Inf}_H^G$ and $\mathrm{Res} = \mathrm{Res}_G^H$. Then*

$$\mathcal{E} \otimes \mathrm{Inf}\,\mathcal{F} \cong_G \mathrm{Inf}(\mathrm{Res}(\mathcal{E}) \otimes \mathcal{F}) \quad , \quad \mathcal{H}om(\mathcal{E}, \mathrm{Inf}\,\mathcal{F}) \cong_G \mathrm{Inf}(\mathcal{H}om(\mathrm{Res}\,\mathcal{E}, \mathcal{F}))\,,$$
$$\mathcal{H}om(\mathrm{Inf}\,\mathcal{F}, \mathcal{E}) \cong_G \mathrm{Inf}(\mathcal{H}om(\mathcal{F}, \mathrm{Res}\,\mathcal{E}))\,.$$

Proof. The underlying sheaf of $\mathcal{E} \otimes \mathrm{Inf}\,\mathcal{F}$ is $\oplus_{[g] \in H \backslash G} \mathcal{E} \otimes g^*\mathcal{F}$. For $\sigma \in G$ the σ-linearization of $\mathcal{E} \otimes \mathrm{Inf}\,\mathcal{F}$ maps $\mathcal{E} \otimes \hat{g}^*F$ isomorphic to $\sigma^*(\mathcal{E} \otimes g^*\mathcal{F})$, where $[g]$ is the image of $[\hat{g}]$ under the action of σ on $H \backslash G$ defined in remark 1.4.2. Furthermore the H-linearization on the component $\mathcal{E} \otimes \mathcal{F}$ coincides with the H-linearization of $\mathrm{Res}(\mathcal{E}) \otimes \mathcal{F}$. By lemma 1.4.1 this gives the result. The other two cases are proven similarly. $\qquad\square$

Also the equivariant pull-backs and push-forwards commute with the inflation functor.

Remark 1.4.4. Let X be a G-scheme and $\pi\colon X \to Y$ a G-invariant morphism, e.g. the quotient morphism. Then there is the composition

$$\pi^* \circ \mathrm{triv} \colon \mathrm{QCoh}(Y) \to \mathrm{QCoh}_G(X) \quad , \quad F \mapsto (\pi^* F, \lambda) \, .$$

For $\sigma \in G$ we denote λ_σ in this case also by σ_*. The map $\sigma_*(U) \colon \pi^* F(U) \to \pi^* F(\sigma(U))$ is given for $s \in (\pi^* F)(U)$ by $(\sigma_*(U)s)(x) = s(\sigma^{-1}x)$ for every ring R and every R-valued point x of $\sigma(U)$. In the later sections, mostly F will be a locally free sheaf and X a variety over $\mathbb{C}$. In this case it suffices to consider $\mathbb{C}$-valued points $x \in \sigma(U)$.

1.4.5 Derived equivariant categories

Let from now on X be a noetherian scheme. We denote the derived categories as

$$\mathrm{D}_G(\mathrm{QCoh}(X)) := \mathrm{D}(\mathrm{QCoh}_G(X)) \quad , \quad \mathrm{D}_G(X) := \mathrm{D}_G(\mathrm{Coh}(X)) := \mathrm{D}(\mathrm{Coh}_G(X)) \, .$$

As usual for $* = +, -, b$ we denote by $\mathrm{D}_G^*(\mathrm{QCoh}(X))$ respectively $\mathrm{D}_G^*(X)$ the full subcategories of complexes with bounded from below, bounded from above or bounded cohomology. For the same reasons as in the non-equivariant case (see e.g. [Huy06, Prop. 3.5]) the category $\mathrm{D}_G^*(X)$ can be identified with the full subcategory of $\mathrm{D}_G^*(\mathrm{QCoh}(X))$ of complexes with coherent cohomology sheaves. Since they are exact, the functors Inf, Res, and in case of the trivial G-action also triv from above define functors on the derived categories. If the scheme X equipped with the trivial G-action is defined over a field k of characteristic zero, the functor of taking invariants also is defined between the derived categories. For an k-scheme X with an arbitrary G-action let $\pi \colon X \to Y$ be a G-invariant morphism of k-schemes, i.e. an equivariant morphism when considering Y with the trivial action. Then we can push forward G-sheaves on X along π and take the G-invariants on Y afterwards. If π_* is an exact functor (which is independent of considering the equivariant functor $\mathrm{QCoh}_G(X) \to \mathrm{QCoh}_G(Y)$ or the non-equivariant one $\mathrm{QCoh}(X) \to \mathrm{QCoh}(Y)$ because of the compatibility with the forgetful functor), we write for short $[\_]^G$ instead of $[\_]^G \circ \pi_*$, which yields also directly a functor on the derived categories.

1.4.6 Injective and locally free sheaves

There are always enough injectives in the category $\mathrm{QCoh}_G(X)$ (see [Gro57, section 5.1]).

Lemma 1.4.5. *Whenever a G-sheaf $(F, \lambda) \in \mathrm{QCoh}_G(X)$ is an injective object also its underlying sheaf $F \in \mathrm{QCoh}(X)$ is injective. If X is defined over a field of characteristic zero also the converse holds.*

Proof. Let $(F, \lambda) \in \mathrm{QCoh}(X)$ be injective and

$$0 \to E' \to E \to E'' \to 0$$

22

a short exact sequence in $\mathrm{QCoh}(X)$. Applying Inf to this sequence yields a short exact sequence in $\mathrm{QCoh}^G(X)$. By the adjointness of Inf and For we get the following isomorphism of complexes:

$$
\begin{array}{ccccccccc}
0 & \longleftarrow & \mathrm{Hom}(E',F) & \longleftarrow & \mathrm{Hom}(E,F) & \longleftarrow & \mathrm{Hom}(E'',F) & \longleftarrow & 0 \\
 & & \downarrow \cong & & \downarrow \cong & & \downarrow \cong & & \\
0 & \longleftarrow & G\,\mathrm{Hom}(\mathrm{Inf}\,E',F) & \longleftarrow & G\,\mathrm{Hom}(\mathrm{Inf}\,E,F) & \longleftarrow & G\,\mathrm{Hom}(\mathrm{Inf}\,E'',F) & \longleftarrow & 0\,.
\end{array}
$$

Since the lower complex is exact by the injectivity of F as an G-sheaf, the upper one is also. This shows the injectivity of $F \in \mathrm{QCoh}(X)$. Taking invariants of G-representations over a field of characteristic zero is exact. Since $G\,\mathrm{Hom}(E,F)$ is defined as the space of invariants of $\mathrm{Hom}(E,F)$ we get the second statement. $\qquad\square$

A *G-equivariant locally free sheaf* is just a coherent G-sheaf whose underlying ordinary sheaf is locally free.

Lemma 1.4.6. *There are enough G-equivariant locally free sheaves in $\mathrm{Coh}_G(X)$ if and only if there are enough locally free sheaves in $\mathrm{Coh}(X)$*

Proof. Let there be enough locally free G-sheaves in $\mathrm{Coh}_G(X)$ and let $F \in \mathrm{Coh}(X)$. Then there is a locally free G-sheaf E and a surjection $\varphi\colon E \to \mathrm{Inf}(F)$. The component $\varphi_e\colon E \to F$ is then a surjection in $\mathrm{Coh}(X)$. Let on the other hand $\mathrm{Coh}(X)$ have enough locally free sheaves and consider $F \in \mathrm{Coh}_G(X)$. Then there is a surjection $\psi\colon E \to \mathrm{For}(F)$ from a locally free sheaf $E \in \mathrm{Coh}(X)$. By the adjoint property of the inflation functor there is a map $\mathrm{Inf}(E) \to F$ in $\mathrm{Coh}_G(X)$. By construction it is still surjective. $\qquad\square$

1.4.7 Derived equivariant functors

Because of the existence of enough injective equivariant sheaves, the functors $\Gamma(X,\_)$, f_* and $\mathcal{H}om$ from subsection 1.4.4 can be derived to (bi-)functors

$$
R\,\Gamma(X,\_)\colon \mathrm{D}_G^+(\mathrm{QCoh}(X)) \to \mathrm{D}^+(\mathrm{Mod}(\mathbb{Z}[G])),\ Rf_*\colon \mathrm{D}_G^+(\mathrm{QCoh}(X)) \to \mathrm{D}_G^+(\mathrm{QCoh}(Y)),
$$
$$
R\,\mathcal{H}om\colon \mathrm{D}_G^-(\mathrm{QCoh}(X))^\circ \times \mathrm{D}_G^+(\mathrm{QCoh}(X)) \to \mathrm{D}_G^+(\mathrm{QCoh}(X))\,.
$$

If there are enough locally free coherent sheaves on Y also the derived equivariant functors

$$
Lf^*\colon \mathrm{D}_G^-(Y) \to \mathrm{D}_G^-(X)\quad,\quad \_\otimes^L\_\colon \mathrm{D}_G^-(Y) \times \mathrm{D}_G^-(Y) \to \mathrm{D}_G^-(Y)
$$

exist. Since the forgetful functor maps injective to injective and locally free to locally free sheaves, all the equivariant derived functors are compatible with their non-equivariant versions via the forgetful functor, for example the following diagram commutes

$$
\begin{array}{ccc}
\mathrm{D}_G^+(\mathrm{QCoh}(X)) & \xrightarrow{\;Rf_*\;} & \mathrm{D}_G^+(\mathrm{QCoh}(Y)) \\
{\scriptstyle\mathrm{For}}\downarrow & & \downarrow{\scriptstyle\mathrm{For}} \\
\mathrm{D}^+(\mathrm{QCoh}(X)) & \xrightarrow[\;Rf_*\;]{} & \mathrm{D}^+(\mathrm{QCoh}(Y))\,.
\end{array}
$$

This implies that (when there are enough locally free sheaves) a derived geometric equivariant functor restricts to a functor between the bounded derived categories of coherent sheaves if and only if the corresponding non-equivariant functor does. Also the functor $G\,\mathrm{Hom}(\_,\_)$ of global G-homomorphisms can be derived. We define $G\,\mathrm{Ext}^i(\_,\_)$ to be the i-th derived functor of $G\,\mathrm{Hom}(\_,\_)$. It coincides with $\mathrm{Hom}_{\mathrm{D}(\mathrm{QCoh}^G(X))}(\_,\_[-i])$. All the common formulas (see e.g. the section "Compatibilities" in [Huy06, chapter 3]) relating the geometric derived functors generalise directly to the equivariant case. In the following lemma we proof one of them as an example.

Lemma 1.4.7. *Let X be a scheme with a G-action such that there are enough locally free sheaves on X and the geometric derived bifunctors restrict to*

$$
R\,\mathcal{H}om,\,\otimes\colon\ \mathrm{D}_G^b(X)\times\mathrm{D}_G^b(X)\to\mathrm{D}_G^b(X)\,.
$$

Then for every $\mathcal{D}^\bullet,\mathcal{E}^\bullet,\mathcal{F}^\bullet\in\mathrm{D}_G^b(X)$ there is a natural isomorphism

$$
R\,\mathcal{H}om(\mathcal{D}^\bullet,\mathcal{E}^\bullet)\otimes^L\mathcal{F}^\bullet\simeq R\,\mathcal{H}om(\mathcal{D}^\bullet,\mathcal{E}^\bullet\otimes^L\mathcal{F}^\bullet)\,.
$$

Proof. Both sides are computed by taking locally free resolutions and then applying the non-derived functors. Thus it suffices to show the formula for locally free equivariant sheaves and the non-derived functors. Let (D,λ), (E,μ) and (F,ν) be G-equivariant locally free sheaves on X. It is well known that the map of ordinary sheaves

$$
T\colon\ \mathcal{H}om(D,E)\otimes F\to\mathcal{H}om(D,E\otimes F)\ ,\quad \varphi\otimes s\mapsto(d\mapsto\varphi(d)\otimes s)
$$

is an isomorphism. We only have to show that it is equivariant. We denote the linearization of the left-hand side by α and that of the right-hand side by β. Then for $g\in G$ indeed

$$
\begin{aligned}
(g^*T)(\alpha_g(\varphi\otimes s)) &= (g^*T)((\mu_g\circ\varphi\circ\lambda_g^{-1})\otimes\nu_g(s)) = \big(d\mapsto(\mu_g\circ\varphi\circ\lambda_g^{-1})(d)\otimes\nu_g(s)\big)\\
&= \beta_g(d\mapsto\varphi(d)\otimes s)\\
&= \beta_g(T(\varphi\otimes s))\,.
\end{aligned}
$$

$\square$

1.4.8 Representations as G-sheaves

Let G be a finite group, V a representation of G over a field k. Let G act on a k-scheme $p\colon X \to \operatorname{Spec} k$. Then applying the equivariant pull-back we obtain the G-sheaf p^*V on X. Its underlying ordinary sheaf is $\mathcal{O}_X^{\dim V}$. For $\mathcal{F}^\bullet \in \mathrm{D}_G^b(X)$ we will write

$$\mathcal{F}^\bullet \otimes V := \mathcal{F}^\bullet \otimes_k V := \mathcal{F}^\bullet \otimes_{\mathcal{O}_X} p^*V.$$

For G-equivariant morphisms $f\colon Y \to X$ and $g\colon X \to Z$ such that the derived pull-backs respectively push-forwards exist we have the isomorphisms $Lf^*(\mathcal{F}^\bullet \otimes V) \simeq_G (Lf^*\mathcal{F}^\bullet) \otimes V$ and $Rg_*(\mathcal{F}^\bullet \otimes V) \simeq_G (Rg_*\mathcal{F}^\bullet) \otimes V$. The second isomorphism is due to the (equivariant) projection formula. Let I be a finite set. The *alternating representation* $\mathfrak{a} = \mathfrak{a}_I$ of $\mathfrak{S}_I$ is the one dimensional vector space with $\sigma \in \mathfrak{S}_I$ acting as multiplication by $\operatorname{sgn}(\sigma)$. Let $\mathfrak{S}_I$ act trivially on a k-scheme X and let $(F, \lambda) \in \operatorname{QCoh}_{\mathfrak{S}_I}(X)$. Then the *sheaf of anti-invariants* $F^{-\mathfrak{S}_I}$ is given by

$$F^{-\mathfrak{S}_I}(U) := \{s \in F(U) \mid \lambda_\sigma(s) = \operatorname{sgn}(\sigma) \cdot s \,\forall\, \sigma \in \mathfrak{S}_I\}.$$

We have $F^{-\mathfrak{S}_n} = (F \otimes \mathfrak{a})^{\mathfrak{S}_I}$.

1.4.9 Equivariant Grothendieck duality

Also Grothendieck duality generalises to proper G-equivariant morphisms (see [LH09]).

Theorem 1.4.8. *Let the finite group G act on schemes X and Y which are of finite type over $\mathbb{C}$ and let $f\colon X \to Y$ be a proper G-equivariant morphism. Then there exists an exact functor $f^!\colon \mathrm{D}^+(\operatorname{QCoh}^G(Y)) \to \mathrm{D}^+(\operatorname{QCoh}^G(X))$ which is compatible with the non-equivariant twisted inverse image functor via the forgetful functor and has the property that for every $\mathcal{F}^\bullet \in \mathrm{D}^-(\operatorname{QCoh}^G(X))$ and $\mathcal{G}^\bullet \in \mathrm{D}^+(\operatorname{QCoh}^G(Y))$ the following holds in $\mathrm{D}(\operatorname{QCoh}^G(Y))$:*

$$Rf_* R\mathcal{H}om_Y(\mathcal{F}^\bullet, f^!\mathcal{G}^\bullet) \simeq R\mathcal{H}om_X(Rf_*\mathcal{F}^\bullet, \mathcal{G}^\bullet).$$

We will need the case where $f = \iota\colon Z \to X$ is a regular embedding with vanishing ideal $\mathcal{I}_Z$ of a G-invariant subvariety. In this case the normal sheaf $N_Z := (\iota^*\mathcal{I}_Z)^\vee$ is a locally free sheaf of rank equal to the codimension of Z in X. If G acts on a scheme X, the structural sheaf always has a natural G-linearization given by $\lambda_g = (g^\#)^{-1}$, where $g^\#\colon g^*\mathcal{O}_X \to \mathcal{O}_X$ is the map of sheaves which belongs to the morphism $g \in \operatorname{Aut}(X)$. Now if Z is a G-invariant closed subscheme with vanishing ideal $\mathcal{I}$, the map $g^\#$ restricts to $g^\#\colon g^*\mathcal{I} \to \mathcal{I}$. Thus the sheaf $\mathcal{I}$ carries a natural G-linearization λ. In the following we will always consider $\mathcal{I}_Z$ and N_Z equipped with the linearization induced by λ.

Proposition 1.4.9. *Let X be a variety over a field k and $\iota\colon Z \to X$ be the immersion of a closed G-invariant local complete intersection subvariety of codimension c with vanishing ideal $\mathcal{I}$ and $\mathcal{G}^\bullet \in \mathrm{D}^b_G(\mathrm{QCoh}(X))$. Then there are canonical G-equivariant isomorphisms*

(i)

$$(\iota_*\mathcal{O}_Z)^\vee \simeq \iota_*(\wedge^c N_Z)[-c]\,,$$

(ii)

$$\mathcal{E}xt^k(\iota_*\mathcal{O}_Z, \iota_*\mathcal{O}_Z) \cong \iota_*(\wedge^k N_Z) \quad \forall\, 0 \le k \le c\,,$$

(iii)

$$\iota^!\mathcal{G}^\bullet \simeq L\iota^*\mathcal{G}^\bullet \otimes (\wedge^c N_Z)[-c]\,.$$

That means, that the G-linearizations on the right sides of the formulas are all the ones canonically induced by the linearization λ of $\mathcal{I}$.

Proof. The proposition is proved in chapter 28 of [LH09] in the more general framework of diagrams of schemes. How to obtain schemes with a group action as a special case is explained in the introduction and chapter 29. $\qquad\square$

1.5 Preliminary lemmas

1.5.1 Derived bifunctors

Let $\mathcal{A}$, $\mathcal{B}$ and $\mathcal{C}$ be abelian categories and $F\colon \mathcal{A} \times \mathcal{B} \to \mathcal{C}$ be an additive bifunctor which is left exact in both arguments. The functor $K^+(F)\colon K^+(\mathcal{A}) \times K^+(\mathcal{B}) \to K^+(\mathcal{C})$ is defined by

$$K(F)(A^\bullet, B^\bullet) := F^\bullet(A^\bullet, B^\bullet) := \mathrm{tot}\, F(A^\bullet, B^\bullet)\,.$$

We assume that there is a full additive subcategory $\mathcal{J}$ of $\mathcal{B}$ such that for every $B \in \mathcal{J}$ the functor $F(\_, B)\colon \mathcal{A} \to \mathcal{C}$ is exact and for every $A \in \mathcal{A}$ the subcategory $\mathcal{J}$ is a $F(A, \_)$-adapted class. Under these assumptions the right derived bifunctor

$$RF\colon \mathrm{D}^+(\mathcal{A}) \times \mathrm{D}^+(\mathcal{B}) \to \mathrm{D}^+(\mathcal{C})$$

exists. Furthermore $K^+(\mathcal{J})$ is a $K^+(F)(A^\bullet, \_)$-adapted class for every $A^\bullet \in K^+(\mathcal{A})$. Thus also the right derived functor $R(K^+(F)(A^\bullet, \_))$ exists and we have for each $A^\bullet \in \mathcal{A}$ and $B^\bullet \in \mathcal{B}$

$$RF(A^\bullet, B^\bullet) \simeq R(K^+(F)(A^\bullet, \_))(B^\bullet)$$

(see [KS06, Section 13.4]). An example were the above assumptions are fulfilled is for any scheme X the functor

$$\mathcal{H}om\colon \mathrm{QCoh}(X)^{\circ} \times \mathrm{QCoh}(X) \to \mathrm{QCoh}(X),$$

where we can choose $\mathcal{J}$ as the class of all injective sheaves (see [Har66, Lemma II 3.1]).

Proposition 1.5.1. *Under the assumptions from above let $A^{\bullet} \in \mathrm{D}^{+}(\mathcal{A})$ and $B^{\bullet} \in \mathrm{D}^{+}(\mathcal{B})$ be complexes such that $R^q F(A^i, B^j) = 0$ for all $q \neq 0$ and all pairs $i, j \in \mathbb{Z}$. Then we have $RF(A^{\bullet}, B^{\bullet}) \simeq F^{\bullet}(A^{\bullet}, B^{\bullet})$.*

Proof. We show that we can enlarge the $K^{+}(F)(A^{\bullet}, \_)$-adapted subcategory $K^{+}(\mathcal{J})$ to the $K^{+}(F)(A^{\bullet}, \_)$-adapted subcategory $K^{+}(\mathcal{J}')$ consisting of all complexes $B^{\bullet}$ with the property as above, i.e. $\mathcal{J}'$ is the full subcategory of all objects $B \in \mathcal{B}$ which are $F(A^i, \_)$-acyclic for every $i \in \mathbb{Z}$. The subcategory $\mathcal{J}'$ is $F(A^i, \_)$-adapted for every $i \in \mathbb{Z}$ (KS lemma 13.3.12). Thus for every acyclic complex $B^{\bullet} \in K^{+}(\mathcal{J}')$ the double complex $F(A^{\bullet}, B^{\bullet})$ has exact columns. Using the spectral sequence

$$E_2^{i,j} = \mathcal{H}_I^i(\mathcal{H}_{II}^j(F(A^{\bullet}, B^{\bullet}))) \implies \mathcal{H}^n(\mathrm{tot}(F(A^{\bullet}, B^{\bullet}))) = \mathcal{H}^n(K^{+}(F)(A^{\bullet}, B^{\bullet}))$$

we see that $K^{+}(F)(A^{\bullet}, B^{\bullet})$ is again acyclic. Hence the category $K^{+}(\mathcal{J}')$ is indeed adapted to the functor $K^{+}(F)(A^{\bullet}, \_)$ and we can use it to compute the derived functor. We get for $B^{\bullet} \in K^{+}(\mathcal{J}')$ as desired

$$RF(A^{\bullet}, B^{\bullet}) = R(K^{+}(F)(A^{\bullet}, \_))(B^{\bullet}) = F^{\bullet}(A^{\bullet}, B^{\bullet}).$$

$\square$

Clearly there is an analogous statement for bifunctors which are right exact in each variable. For a fixed object $A^{\bullet} \in \mathrm{D}^{+}(\mathcal{A})$ and F as above, $G := K^{+}(F)(A^{\bullet}, \_)$ and $\mathcal{J}'$ as in the proof we call every $B \in \mathcal{J}'$ a *G-acyclic* object.

1.5.2 Danila's lemma and corollaries

Let G be a finite group acting transitively on a finite set I, R a ring, and M a $R[G]$-module admitting a decomposition $M = \oplus_{i \in I} M_i$ such that for any $i \in I$ and $g \in G$ the action of g on M restricted to M_i is an isomorphism $g\colon M_i \xrightarrow{\cong} M_{g(i)}$. Then the G-action on M induces a $\mathrm{Stab}_G(i)$-action on M_i, which makes the projection $M \to M_i$ a $\mathrm{Stab}_G(i)$-equivariant map.

Lemma 1.5.2 ([Dan01]). *For all $i \in I$ the projection $M \to M_i$ induces an isomorphism $M^G \xrightarrow{\cong} M_i^{\mathrm{Stab}_G(i)}$.*

Proof. The inverse is given by $m_i \mapsto \oplus_{[g]\in G/\operatorname{Stab}_G(i)} g \cdot m_i$ with $g \cdot m_i \in M_{g(i)}$. $\qquad\square$

We can globalise Danila's lemma to G-sheaves. Let G and I be as above and G act on a scheme X. Let $\mathcal{M}$ be a G-sheaf on X admitting a decomposition $\mathcal{M} = \oplus_{i\in I}\mathcal{M}_i$ such that for any $i \in I$ and $g \in G$ the linearization λ restricted to $\mathcal{M}_i$ is an isomorphism $\lambda_g \colon \mathcal{M}_i \xrightarrow{\cong} g^*\mathcal{M}_{g(i)}$. Then the G-linearization of $\mathcal{M}$ restricts to a $\operatorname{Stab}_G(i)$-linearization of $\mathcal{M}_i$, which makes the projection $\mathcal{M} \to \mathcal{M}_i$ a $\operatorname{Stab}_G(i)$-equivariant morphism. By lemma 1.4.1 for every $i \in I$ we have

$$\mathcal{M} \cong_G \operatorname{Inf}^G_{\operatorname{Stab}(i)} \mathcal{M}_i \, .$$

Corollary 1.5.3. *Let $\pi \colon X \to Y$ be a G-invariant morphism of schemes. Then for all $i \in I$ the projection $\mathcal{M} \to \mathcal{M}_i$ induces an isomorphism $(\pi_*\mathcal{M})^G \xrightarrow{\cong} (\pi_*\mathcal{M}_i)^{\operatorname{Stab}_G(i)}$.*

Proof. For an affine open $U \subset Y$ we set $R = \mathcal{O}(U)$, $M = \Gamma(\pi^{-1}U, \mathcal{M})$ and $M_i = \Gamma(\pi^{-1}U, \mathcal{M}_i)$. Then the lemma applies and gives the isomorphism of sheaves over U. Since for varying U all the isomorphisms are induced by the projection to the i-th summand, they glue together. $\quad\square$

Remark 1.5.4. The assertion of the lemma respectively the corollary remains true if we consider complexes of $R[G]$-modules $M^\bullet$ and $M_i^\bullet$ respectively complexes of G-sheaves. Let k be a field of characteristic zero. If R is a k-algebra respectively in the case of G-sheaves if X and Y are k-schemes and π_* is exact, taking invariants is exact. Hence in this case we also have Danila's lemma for the cohomology of the complexes, i.e.

$$[\mathcal{H}^k(M^\bullet)]^G \cong [\mathcal{H}^k(M_i^\bullet)]^{\operatorname{Stab}_G(i)} \, .$$

Let G act on a scheme X and let $\mathcal{E}^\bullet \in \operatorname{D}^b_G(X)$. Let F be one of the functors $\mathcal{E}^\bullet \otimes \_$, $\mathcal{H}om(\mathcal{E}^\bullet, \_)$ or $\mathcal{H}om(\_, \mathcal{E}^\bullet)$. We assume that there are enough locally free sheaves on X. Then the functor F can be derived and we denote its right respectively left derived by DF. Let $H \leq G$ be a subgroup. We can consider F and DF as functors on the H-equivariant categories by replacing $\mathcal{E}^\bullet$ by $\operatorname{Res}^H_G \mathcal{E}^\bullet$.

Lemma 1.5.5. *For $\mathcal{F}^\bullet \in \operatorname{D}^b_H(X)$ there is in $\operatorname{D}_G(\operatorname{QCoh}(X))$ a natural isomorphism*

$$DF(\operatorname{Inf}^G_H \mathcal{F}^\bullet) \simeq \operatorname{Inf}^G_H(DF(\mathcal{F}^\bullet)) \, .$$

Proof. Let $\mathcal{A}^\bullet$ be a F-acyclic H-equivariant resolution of $\mathcal{F}^\bullet$. Then $DF(\mathcal{F}^\bullet) \simeq F(\mathcal{A}^\bullet)$ in $\operatorname{D}_H(\operatorname{QCoh}(X))$. Since the inflation functor is exact $\operatorname{Inf}(\mathcal{A}^\bullet)$ is a G-equivariant resolution of $\operatorname{Inf} \mathcal{F}^\bullet$. The objects $\operatorname{Inf}(\mathcal{A}^i)$ are still F-acyclic because of the compatibility of DF with the forgetful functor and with direct sums. Thus, $DF(\operatorname{Inf} \mathcal{F}^\bullet) \simeq F(\operatorname{Inf}(\mathcal{A}^\bullet)) \simeq \operatorname{Inf}(F(\mathcal{A}^\bullet))$ holds using lemma 1.4.3. $\qquad\square$

Corollary 1.5.6. *Let G, I, X, $\pi\colon X \to Y$, $\mathcal{M}$ and F be as above such that X and Y are schemes over a field of characteristic zero and π_* is exact. Then there are natural isomorphisms $[\pi_* DF(\mathcal{M})]^G \simeq [\pi_* DF(\mathcal{M}_i)]^{\mathrm{Stab}(i)}$ and $[\pi_* D^k F(\mathcal{M})]^G \cong [\pi_* D^k F(\mathcal{M}_i)]^{\mathrm{Stab}(i)}$ for every $k \in \mathbb{Z}$.*

Proof. Using the identification of $\mathcal{M}$ with the inflation of $\mathcal{M}_i$ we can conclude by the previous lemma and the previous remark. □

We can get analogous results with F being the push-forward or the pull-back along an equivariant morphism.

Remark 1.5.7. Let G, I and M be as above, $N = \oplus_{j \in J} N_j$ a further $R[G]$ module such that G acts transitive on J and such that $g\colon N_j \to N_{g(j)}$ for all $j \in J$. Let $\varphi\colon M \to N$ be a morphism of $R[G]$-modules with components $\varphi(i,j)\colon M_i \to N_j$. Then for fixed $i \in I$ and $j \in J$ the map φ^G under the isomorphisms $M^G \cong M_i^{\mathrm{Stab}(i)}$ and $N^G \cong N_j^{\mathrm{Stab}(j)}$ of lemma 1.5.2 is given by (see also [Sca09a, Appendix B])

$$\varphi^G \colon M_i^{\mathrm{Stab}(i)} \to N_j^{\mathrm{Stab}(j)} \quad , \quad m \mapsto \sum_{[g] \in \mathrm{Stab}(i)\backslash G} \varphi(g(i),j)(g \cdot m) \,.$$

Clearly, there is the analogous formula in the case of G-sheaves.

Remark 1.5.8. Danila's lemma and the corollaries can also be used to simplify the computation of invariants if G does not act transitively on I. In that case let $I_1, \ldots, I_k$ be the G-orbits in I. Then G acts transitively on I_ℓ for every $1 \leq \ell \leq k$ and the lemma can be applied to every $M_{I_\ell} = \oplus_{i \in I_\ell} M_i$ instead of M. Choosing representatives $i_\ell \in I_\ell$ yields

$$M^G \cong \bigoplus_{\ell=1}^{k} M_{i_\ell}^{\mathrm{Stab}_G(i_\ell)} \,.$$

Lemma 1.5.9. *Let k be a field of characteristic zero, R a k-algebra, G a finite group and M an $R[G]$-module. Let N be a R-module, i.e. a $R[G]$-module where G is acting trivially. Then*

$$(M \otimes_R^L N)^G = M^G \otimes_R^L N \,.$$

Proof. See [Sca09a, Lemma 1.7.1]. □

Also this lemma can be globalised to get an analogous result for G-sheaves.

1.5.3 Pull-back along regular embeddings

Let G be a finite group. In this subsection every variety is supposed to be a G-variety (over a fixed field k) and every subvariety is G-invariant and considered with the restricted G-action, i.e. all embeddings are G-equivariant. Also, all sheaves and the considered functors and derived functors are equivariant. Of course one can apply the results to the non-equivariant case by setting $G = 1$. Throughout the whole section X is a non-singular variety.

Lemma 1.5.10. *Let* $j \colon Z \to X$ *be an embedding of varieties. Then* $j_* \colon \mathrm{Coh}_G(Z) \to \mathrm{Coh}_G(X)$ *is a fully faithful functor.*

Proof. As in the non-equivariant case for $E, F \in \mathrm{Coh}_G(X)$ there are the natural isomorphisms

$$G \operatorname{Hom}_X(j_*E, j_*F) \cong G \operatorname{Hom}_Z(j^*j_*E, F) \cong G \operatorname{Hom}_Z(E, F).$$

$\square$

Lemma 1.5.11 ([Sca09a, Lemma A.2])**.** *Let* A *be a regular notherian local ring,* $M_1, \dots, M_\ell$ *finite Cohen-Macauley modules over* A, *such that*

$$\mathrm{codim}(M_1 \otimes \cdots \otimes M_\ell) = \sum_{i=1}^{\ell} \mathrm{codim}(M_i).$$

Then all the higher torsion modules vanish, i.e. $\mathrm{Tor}_i^A(M_1, \dots, M_\ell) = 0$ *for all* $i > 0$.

Corollary 1.5.12. *Let* $i \colon Y \hookrightarrow X$ *and* $j \colon Z \hookrightarrow X$ *be embeddings of Cohen-Macaulay subvarieties which intersect properly in* X, *i.e.*

$$\mathrm{codim}(Y \cap Z) = \mathrm{codim}(Y) + \mathrm{codim}(Z),$$

and F *a Cohen-Macaulay sheaf on* Z. *Then all the derived pull-backs* $L^{-q}i^*j_*F$ *for* $q > 0$ *vanish.*

Proof. By projection formula $i_*L^{-q}i^*j_*F \cong \mathcal{T}or_q^{\mathcal{O}_X}(i_*\mathcal{O}_Y, j_*F)$. Since $\mathcal{O}_Y$ and F are Cohen-Macaulay, $i_*\mathcal{O}_Y$ and j_*F are also (see [Ser00, IV B prop 11]). The stalks of the higher torsion sheaves can be computed as the torsion modules of the stalks so the result follows from the two previous lemmas. $\square$

Lemma 1.5.13. *(i) Let* $f \colon X \to Y$ *be a* G-*equivariant morphism such that the derived pull-back exists and* $\mathcal{E}^\bullet \in \mathrm{D}_G^-(Y)$ *a complex such that the cohomology* $\mathcal{H}^q(\mathcal{E}^\bullet)$ *is* f^*-*acyclic for all* $q \in \mathbb{Z}$. *Then* $L^q f^* \mathcal{E}^\bullet \cong f^* \mathcal{H}^q(\mathcal{E}^\bullet)$ *for all* $q \in \mathbb{Z}$.

(ii) Let $\mathcal{F} \in \mathrm{Coh}_G(X)$ *such that* $\mathcal{H}^q(\mathcal{E}^\bullet)$ *is* $(\mathcal{F} \otimes \_)$-*acyclic for all integers* q. *Then* $\mathcal{T}or_q(\mathcal{F}, \mathcal{E}^\bullet) \cong \mathcal{F} \otimes \mathcal{H}^{-q}(\mathcal{E}^\bullet)$ *for all* $q \in \mathbb{Z}$

Proof. We consider the spectral sequence

$$E_2^{p,q} = L^p f^*(\mathcal{H}^q(\mathcal{E}^\bullet)) \Longrightarrow E^n = L^n f^*(\mathcal{E}^\bullet)$$

(see [Huy06, p.81]). By the assumption this spectral sequence is concentrated on the q-axis, hence $E^q = E_2^{0,q}$ for each integer q. For the second part we use the spectral sequence

$$E_2^{p,q} = \mathcal{T}or_{-p}(\mathcal{F}, \mathcal{H}^q(\mathcal{E}^\bullet)) \Longrightarrow E^n = \mathcal{T}or_{-n}(\mathcal{F}, \mathcal{E}^\bullet).$$

$\square$

Let S be a variety and Y, Z closed subvarieties. We denote the inclusions by

$$\begin{array}{ccc} Y \cap Z & \xrightarrow{\ d\ } & Y \\ {\scriptstyle c}\downarrow & & \downarrow{\scriptstyle b} \\ Z & \xrightarrow{\ \ a\ \ } & S. \end{array}$$

Lemma 1.5.14. *For every $F \in \mathrm{QCoh}_G(Z)$ the base change formula $b^* a_* F \cong d_* c^* F$ holds.*

Proof. We reduce to the case $S = \operatorname{Spec} A$ affine and notice that $\frac{M}{I_Y \cdot M} = \frac{M}{(I_Y + I_Z) \cdot M}$ holds for every A/I_Z-module M. $\square$

Lemma 1.5.15. *Let $j \colon S \hookrightarrow X$ be a regular embedding of codimension c with vanishing ideal $\mathcal{I}_S$ and $F \in \mathrm{Coh}_G(X)$. Then for every $q \in \mathbb{Z}$*

$$L^{-q} j^* j_* F \cong F \otimes \wedge^q(j^* \mathcal{I}_S).$$

Proof. We first proof the special case $F = \mathcal{O}_S$. By 1.4.9 (iii)

$$j_*(Lj^* j_* \mathcal{O}_S \otimes \wedge^c j^* \mathcal{I}_S^\vee[-c]) \simeq j_* j^! j_* \mathcal{O}_S \simeq j_* R\mathcal{H}om_S(\mathcal{O}_S, j^! j_* \mathcal{O}_S) \simeq R\mathcal{H}om_X(j_* \mathcal{O}_S, j_* \mathcal{O}_S).$$

Now taking the $(c-q)$-th cohomology on both sides by 1.4.9 (ii) yields

$$j_*(L^{-q} j^* j_* \mathcal{O}_S \otimes \wedge^c j^* \mathcal{I}_S^\vee) \cong j_*(\wedge^{c-q} j^* \mathcal{I}_S^\vee).$$

Using the fact that j_* is fully faithful we can chancel it from the isomorphism. Tensoring with $\wedge^c j^* \mathcal{I}_S$ gives the result. For general F using the projection formula twice we get

$$j_* Lj^* j_* F \simeq j_*(Lj^* j_* F \otimes \mathcal{O}_S) \simeq j_* F \otimes^L j_* \mathcal{O}_S \simeq j_*(F \otimes^L Lj^* j_* \mathcal{O}_S).$$

We have already proven, that the $L^q j^* j_* \mathcal{O}_S$ are locally free, hence $(F \otimes \_)$-acyclic. Thus we can use 1.5.13 (ii) with $\mathcal{E}^\bullet = Lj^* j_* \mathcal{O}_S$ which proves the general formula. $\square$

Let $j \colon S \hookrightarrow X$ and $S' \hookrightarrow S$ be regular embeddings. Then the composition $j' \colon S' \hookrightarrow X$ is also a regular embedding (see e.g. [Ful98, Appendix B.7]). We denote by $\pi \colon j_* \mathcal{O}_S \to j'_* \mathcal{O}_{S'}$ the natural surjection.

Lemma 1.5.16. *The induced morphism of the $(-q)$-th derived pull-back functors*

$$L^{-q}j'_*(\pi)\colon L^{-q}j'^*j_*\mathcal{O}_S \cong \wedge^q(j^*\mathcal{I}_S)_{|S'} \to \wedge^q j'^*\mathcal{I}_{S'} \cong L^{-q}j'^*j'_*\mathcal{O}_{S'}$$

coincides with the one induced by the inclusion $\mathcal{I}_S \subset \mathcal{I}_{S'}$.

Proof. Because of the compatibility of the equivariant derived functors with the non-equivariant ones via the forgetful functor, we can compute $L^{-q}j'^*(\pi)$ on the non-equivariant derived functor. The question is local. Thus we can assume that both vanishing ideals are globally generated by regular sequences

$$\mathcal{I}_S = (f_1,\ldots,f_\alpha) \quad , \quad \mathcal{I}_{S'/S} = (f_{\alpha+1},\ldots,f_{\alpha+\beta}).$$

Now the non-equivariant derived pull-backs are computed (see e.g. [Huy06, chapter 11.1]) using the Koszul complexes $K^\bullet(f_1,\ldots f_\alpha)$ and $K^\bullet(f_1,\ldots f_{\alpha+\beta})$ as free resolutions of $j_*\mathcal{O}_S$ and $j'_*\mathcal{O}_{S'}$. After pulling back the Koszul complexes along j' all the differentials vanish which yields the isomorphism

$$L^{-q}j'^*j_*\mathcal{O}_S \cong \mathcal{H}^{-q}(j'^*K^\bullet(f_i,\ldots,f_\alpha)) \xrightarrow{\cong} \wedge^q j'^*\mathcal{I}_S, \quad e_{i_1} \wedge \cdots \wedge e_{i_q} \mapsto f_{i_1} \wedge \cdots \wedge f_{i_q}$$

and the analogous isomorphism for S' instead of S. The morphism π can be continued on the Koszul resolutions by

$$K^{-q}(f_1,\ldots,f_\alpha) \to K^{-q}(f_1,\ldots,f_{\alpha+\beta}), e_{i_1} \wedge \cdots \wedge e_{i_q} \mapsto e_{i_1} \wedge \cdots \wedge e_{i_q}$$

which yields the result. $\qquad\qquad\square$

The two lemmas 1.5.12 and 1.5.15 can be combined as follows. Let again Y and Z be closed Cohen-Macaulay subvarieties of X. We assume that there is a non-singular subvariety $S \hookrightarrow X$ such that S contains Y and Z such that Y and Z intersect properly in S. Note that the embedding $S \hookrightarrow X$ is regular since both S and X are non-singular. We use the following notations for the closed embeddings:

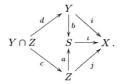

Lemma 1.5.17. *Let $F \in \mathrm{Coh}_G(Z)$ be Cohen-Macauley. With the notations introduced above,*

for $q \in \mathbb{Z}$ there is the following isomorphism

$$L^{-q}i^* j_* F \cong d_* c^* \left(F \otimes \left(a^* \wedge^q \iota^* \mathcal{I}_S\right)\right) .$$

Proof. We have $L^{-q}i^* j_* F \cong L^{-q}b^*(L\iota^* \iota_*(a_* F))$. Also by lemma 1.5.15 and the projection formula

$$L^{-q}\iota^* \iota_*(a_* F) \cong a_* F \otimes \wedge^q(\iota^* \mathcal{I}_S) \cong a_*(F \otimes a^* \wedge^q N_S^\vee) .$$

Since F is Cohen-Macaulay and $\wedge^q N_S^\vee$ is locally free, the whole $F \otimes a^* \wedge^q N_S^\vee$ is Cohen-Macaulay for every q. So by 1.5.12 the assumptions of 1.5.13 (i) are satisfied with $f = b$ and $\mathcal{E}^\bullet = L\iota^* \iota_*(a_* F)$. Thus,

$$L^{-q}i^* j_* F \cong b^*(a_* F \otimes \wedge^q(\iota^* \mathcal{I}_S)) \overset{\text{PF}}{\cong} b^* a_*(F \otimes a^* \wedge^q \iota^* \mathcal{I}_S) \overset{1.5.14}{\cong} d_* c^* \left(F \otimes \left(a^* \wedge^q \iota^* \mathcal{I}_S\right)\right) .$$

$\square$

Let Y, Z and S be as above and $S' \hookrightarrow S$ another non-singular subvariety such that S' contains Y and intersects properly with Z in S. We denote the intersection by $Z' = Z \cap S'$. It is again Cohen-Macaulay (see [Ser00, section IV.B.2]). Furthermore, the subvarieties Z' and Y intersect properly in S'. Hence, the above lemma applies again and

$$L^{-q}i^* j'_* \mathcal{O}_{Z'} \cong \wedge^q(N_{S'}^\vee)_{|Y \cap Z} ,$$

where j' is the inclusion of Z' in X.

Lemma 1.5.18. *Let $\pi \colon j_* \mathcal{O}_Z \to j'_* \mathcal{O}_{Z'}$ be the natural surjection. Then the morphism*

$$L^{-q}i^*(\pi) \colon L^{-q}i^* j_* \mathcal{O}_Z \cong \wedge^q(N_S^\vee)_{|Y \cap Z} \to \wedge^q(N_{S'}^\vee)_{|Y \cap Z'} \cong L^{-q}i^* j'_* \mathcal{O}_{Z'}$$

is the one induced by the inclusion $\mathcal{I}_S \subset \mathcal{I}_{S'}$.

Proof. We have $\mathcal{O}_{Z'} \cong \mathcal{O}_Z \otimes_{\mathcal{O}_S} \mathcal{O}_{S'}$ and the surjection $\pi \colon \mathcal{O}_Z \to \mathcal{O}_{Z'}$ is induced by the surjection $\mathcal{O}_S \to \mathcal{O}_{S'}$. Thus we can use lemma 1.5.16. $\square$

1.5.4 Partial diagonals and the standard representation

Let X be a smooth variety of dimension d over $\mathbb{C}$. Let $I = \{i_1 < \cdots < i_\ell\} \subset [n]$ with $\#I = \ell \geq 2$. Then the partial diagonal Δ_I is a $\text{Stab}_{\mathfrak{S}_n}(I) = \mathfrak{S}_I \times \mathfrak{S}_{\bar{I}}$-equivariant smooth subvariety of codimension $d(\ell - 1)$ in X^n. The subgroup $\mathfrak{S}_I$ acts trivially on Δ_I. Thus, the $\mathfrak{S}_I$-linearization of the conormal bundle $N_I^\vee$ is just a $\mathfrak{S}_I$-action (see subsection 1.4.3). Since $N_I^\vee$ is locally free, this action is determined by the action on the fibers $N_I^\vee(Q) = \mathcal{I}_I(Q)$ over closed points $Q \in \Delta_I$. For $\ell \in \mathbb{N}$ the *natural representation* V_I of $\mathfrak{S}_I$ is the vector space $\mathbb{C}^I$

with $\mathfrak{S}_\ell$ permuting the vectors of the standard base $(e_i \mid i \in I)$. The natural representation decomposes into the direct sum of its invariants, which form the one-dimensional vector space spanned by $\sum_{i \in I} e_i$, and the space $\{v \in V \mid \sum_{i \in I} v_i = 0\}$ spanned by the vectors $e_{i_1} - e_{i_2}, e_{i_2} - e_{i_3}, \ldots, e_{i_{\ell-1}} - e_{i_\ell}$. The latter summand is an irreducible representation of $\mathfrak{S}_I$ called the *standard representation* and denoted by ϱ_I. For $I = [\ell]$ we denote the standard representation of $\mathfrak{S}_I = \mathfrak{S}_\ell$ also by ϱ_ℓ or simply ϱ. We need the following result from local algebra (see [Ser00, IV Proposition 22]).

Proposition 1.5.19. *If $\{x_1, \ldots, x_p\}$ are p elements of the maximal ideal $\mathfrak{m}$ of a regular local ring A, the following three properties are equivalent:*

(i) *$x_1, \ldots, x_p$ is part of a regular system of parameters of A, i.e. of a system of parameters which generates $\mathfrak{m}$.*

(ii) *The images of $x_1, \ldots, x_p$ in $\mathfrak{m}/\mathfrak{m}^2$ are linearly independent over k.*

(iii) *The local ring $A/(x_1, \ldots, x_p)$ is regular and has dimension $\dim A - p$.*
(In particular, $(x_1, \ldots, x_p)$ is a prime ideal.)

Lemma 1.5.20. *Let $I \subset [n]$ and $Q \in \Delta_I$ be a closed point. Considering $\mathcal{I}_I(Q)$ with the natural action of $\mathfrak{S}_I$ (see subsection 1.4.9) we have $\mathcal{I}_I(Q) \cong \varrho_I^{\oplus d}$.*

Proof. For each closed point $P \in X$ we choose an open affine neighbourhood with coordinate ring $A(P)$. We denote the maximal ideal of $A(P)$ corresponding to P by $\mathfrak{m}(P)$, hence $\mathcal{O}_{X,P} = A(P)_{\mathfrak{m}(P)}$. Furthermore we choose a regular system of parameters $x_1(P), \ldots, x_d(P)$ of $\mathcal{O}_{X,P}$. By multiplying them by their denominator if necessary, we may assume that $x_1(P), \ldots, x_d(P) \in \mathfrak{m}(P) \subset A(P)$. Let $Q = (P^1, \ldots, P^n) \in X^n$. Then there is an affine open neighbourhood of Q in the product X^n with coordinate ring $B(Q) = A(P^1) \otimes \cdots \otimes A(P^n)$. The point Q corresponds to the maximal ideal

$$\mathfrak{n}(Q) = \mathfrak{m}(P^1) \otimes A \otimes \cdots \otimes A + A \otimes \mathfrak{m}(P^2) \otimes \cdots \otimes A + \cdots + A \otimes A \otimes \cdots \otimes \mathfrak{m}(P^n).$$

Since the $(x_i(P))_{i=1,\ldots,d}$ generate $\mathfrak{m}(P)$, the family $(x_i^j(Q))_{i=1,\ldots,d}^{j=1,\ldots,n}$ with

$$x_i^j(Q) = 1 \otimes \cdots \otimes 1 \otimes x_i(P^j) \otimes 1 \otimes \cdots \otimes 1$$

generates $\mathfrak{n}(Q)$. As $\mathcal{O}_{X^n,Q}$ is a regular local ring of dimension $d \cdot n$, this family is a regular system of parameters. Without loss of generality we may assume that $I = [\ell]$ and consider a point $Q = (P, \ldots, P, P^{\ell+1}, \ldots, P^n)$ of $\Delta_{[\ell]}$. For $1 \leq i \leq d$ and $1 \leq j \leq \ell - 1$, we set $\zeta_i^j = x_i^j(Q) - x_i^{j+1}(Q)$. Clearly, the ζ_i^j are elements of the stalk of the vanishing ideal $(\mathcal{I}_I)_Q \subset \mathfrak{n}(Q)\mathcal{O}_{X^n,Q}$. We denote their images in $\mathfrak{n}(Q)/\mathfrak{n}(Q)^2$ by $\bar{\zeta}_i^j = \bar{x}_i^j(Q) - \bar{x}_i^{j+1}(Q)$. By

1.5.19 the vectors $\bar{x}_1^1(Q), \ldots, \bar{x}_1^n(Q), \bar{x}_2^1(Q), \ldots, \bar{x}_d^n(Q)$ form a basis of $\mathfrak{n}(Q)/\mathfrak{n}(Q)^2$. Hence, $\bar{\zeta}_1^1, \ldots \bar{\zeta}_d^{\ell-1}$ are $d(\ell-1)$ linearly independent elements in $\mathfrak{n}(Q)/\mathfrak{n}(Q)^2$. Thus, by proposition 1.5.19 the ζ_i^j generate a prime ideal in $\mathcal{O}_{X^n,Q}$ of height $d(\ell-1)$. It is contained in $(\mathcal{I}_I)_Q$ which is also prime of the same height. Thus,

$$(\mathcal{I}_I)_Q = (\{\zeta_i^j \mid 1 \le i \le d, 1 \le j \le \ell-1\})$$

and the fiber $\mathcal{I}_I(Q)$ has $(\bar{\zeta}_i^j)_{i=1,\ldots,d}^{j=1,\ldots,\ell-1}$ as a base. Now the isomorphism $\mathcal{I}_I(Q) \xrightarrow{\cong} \varrho^{\oplus d}$ is given by $\bar{\zeta}_i^j \mapsto (0, \ldots, 0, e_j - e_{j+1}, 0, \ldots, 0)$. □

By proposition 1.4.9 the derived dual of the structural sheaf of Δ_I in X^n is cohomologically concentrated in degree $d(\ell-1)$ with

$$\mathcal{H}^{d(\ell-1)}((\iota_*\mathcal{O}_{\Delta_I})^{\mathbf{v}}) \cong \iota_{I*}(\wedge^{d(\ell-1)}\iota_I^*\mathcal{I}_I)^{\vee}$$

equipped with the $\mathfrak{S}_I \times \mathfrak{S}_{\bar{I}}$-linearization which is induced by the natural linearization of $\mathcal{I}_I$.

Lemma 1.5.21. *The group $\mathfrak{S}_I$ acts on $\mathcal{H}^{d(\ell-1)}(\mathcal{O}_{\Delta_I}^{\mathrm{v}})$ trivially if d is even and alternating if d is odd.*

Proof. The sheaf $(\wedge^{d(\ell-1)}\iota_I^*\mathcal{I}_I)^{\vee}$ is a line bundle on Δ_I. Thus, the action of $\mathfrak{S}_I$ is determined by the action on the fibers, which are isomorphic to the representation $\wedge^{d(\ell-1)}(\varrho^{\oplus d})$ (dualizing can be dropped for one-dimensional representations). It suffices to consider the action of transpositions of neighbours $\tau = (k, k+1)$ since $\mathfrak{S}_I$ is generated by those permutations. As one can compute, each such τ acts as a matrix $\varrho(\tau) \in \mathrm{GL}(\mathbb{C}^{\ell-1})$ with determinant -1. The action of τ on $\wedge^{d(\ell-1)}(\varrho^{\oplus d})$ is given by the determinant of $\varrho(\tau)^{\oplus d}$. Thus the assertion follows. □

In the following we will always consider the case that X is a surface over $\mathbb{C}$, i.e $d = 2$. In that case there is also a formula for the dimension of the invariants of the lower exterior powers of $\varrho^{\oplus d}$.

Lemma 1.5.22. *Let ϱ be the standard representation of the symmetric group $\mathfrak{S}_\ell$ for any $\ell \in \mathbb{N}$. The dimension of the space of invariants of the exterior products of the double direct sum of the standard representation is given by*

$$\dim(\wedge^p(\varrho \oplus \varrho))^{\mathfrak{S}_\ell} = \begin{cases} 1 & \text{if } 0 \le p \le 2(\ell-1) \text{ is even} \\ 0 & \text{else.} \end{cases}$$

Proof. For fixed $\ell \in \mathbb{N}$ the natural representation V decomposes as $V = \varrho \oplus \mathbb{C}$. Thus the representation $\wedge^p(V \oplus V)$ decomposes as

$$\wedge^p(V \oplus V) = \wedge^p(\varrho \oplus \varrho) \oplus (\wedge^{p-1}(\varrho \oplus \varrho))^{\oplus 2} \oplus \wedge^{p-2}(\varrho \oplus \varrho)$$

where occurring negative exterior powers (for $p = 0, 1$) are read as zero. For the dimensions of the spaces of invariants this yields

$$\dim(\wedge^p(V \oplus V))^{\mathfrak{S}_\ell} = \dim(\wedge^p(\varrho \oplus \varrho))^{\mathfrak{S}_\ell} + 2\dim(\wedge^{p-1}(\varrho \oplus \varrho))^{\mathfrak{S}_\ell} + \dim(\wedge^{p-2}(\varrho \oplus \varrho))^{\mathfrak{S}_\ell}.$$

The action of $\mathfrak{S}_\ell$ on $\wedge^0(\varrho \oplus \varrho)$ is by definition trivial so the proposition is true for $p = 0$. We will show in the following that $\dim(\wedge^p(V \oplus V))^{\mathfrak{S}_\ell} = 2$ for all $2(\ell - 1) \geq p \geq 1$. Then the general formula for $\dim(\wedge^p(\varrho \oplus \varrho))^{\mathfrak{S}_\ell}$ follows by induction. The dimension of the invariants of a representation U of a finite group G can be computed using characters by the formula (see [FH], 2.9)

$$\dim U^G = \frac{1}{|G|} \sum_{g \in G} \chi_U(g).$$

Following the notation and the idea of the proof of [FH91, Proposition 3.12] we set for $\sigma \in \mathfrak{S}_\ell$ and $M \subset [\ell]$

$$\{\sigma\}_M := \begin{cases} 0 & \text{, if } \sigma(M) \neq M \\ \operatorname{sgn}(\sigma_{|M}) & \text{, if } \sigma(M) = M. \end{cases}$$

Considering the standard basis of $U = \wedge^p(V \oplus V)$ consisting of the vectors

$$e_M^1 \wedge e_N^2 := e_{m_1}^1 \wedge \cdots \wedge e_{m_s}^1 \wedge e_{n_1}^2 \wedge \cdots \wedge e_{n_t}^2$$

for $M = \{m_1 < \cdots < m_s\}$ and $N = \{n_1 < \cdots < n_t\}$ subsets of $[\ell]$ with $s + t = p$, the character is obtained as

$$\chi_U(\sigma) = \sum_{|M|+|N|=p} \{\sigma\}_M \cdot \{\sigma\}_N.$$

A permutation $\sigma \in \mathfrak{S}_\ell$ mapping M to M and N to N is given by permutations on the four sets $M \setminus N$, $N \setminus M$, $N \cap M$ and $[\ell] \setminus (N \cup M)$ which we will denote by a, b, c and d. This yields

$$\ell! \cdot \dim(U^{\mathfrak{S}_\ell}) = \sum_{\sigma \in \mathfrak{S}_\ell} \sum_{|M|+|N|=p} \{\sigma\}_M \cdot \{\sigma\}_N$$

$$= \sum_{|M|+|N|=p} \sum_{(a,b,c,d)} \operatorname{sgn}(a)(\operatorname{sgn}(c))^2 \operatorname{sgn}(b) \quad (*).$$

For $k \geq 2$ there are the same number of even and odd permutations in $\mathfrak{S}_k$. Hence, the inner sum of $(*)$ vanishes whenever $M \setminus N$ or $N \setminus M$ consists of more than one element. Thus in the case that p is even in order to make a contribution M and N have to be of the form $M = (N \setminus \{i\}) \cup \{j\}$ with $i \in N$ and $j \notin N$ or of the form $N = M$. The number of tuples (M, N) of the first form is $\binom{\ell}{p/2} \cdot \frac{p}{2} \cdot (\ell - \frac{p}{2})$. For a fixed tuple (M, N) of this form c can be choosen out of $\#\mathfrak{S}_{M \cap N} = (\frac{p}{2} - 1)!$ and d out of $\#\mathfrak{S}_{[\ell] \setminus (M \cup N)} = (\ell - \frac{p}{2} - 1)!$ elements. The tuple (M, M) can be chosen in $\binom{\ell}{p/2}$ ways. In this case we have $\#\mathfrak{S}_{M \cap N} = (\frac{p}{2})!$ and $\#\mathfrak{S}_{[\ell] \setminus (M \cup N)} = (\ell - \frac{p}{2})!$. In summary

$$(*) = \binom{\ell}{p/2} \frac{p}{2} \left(\ell - \frac{p}{2}\right) \left(\frac{p}{2} - 1\right)! \left(\ell - \frac{p}{2} - 1\right)! + \binom{\ell}{p/2} \left(\frac{p}{2}\right)! \left(\ell - \frac{p}{2}\right)! = \ell! \cdot 2$$

as required. For p odd we get the same result by a similar computation. See also [Sca09a, Appendix C] for another proof. $\square$

Corollary 1.5.23. *For $p \in \mathbb{N}$ an even number and $I \subset [n]$ with $2(|I| - 1) \geq p$ the sheaf $(\wedge^p N_I^\vee)^{\mathfrak{S}_I} = (\wedge^p \iota_I^* \mathcal{I}_I)^{\mathfrak{S}_I}$ is a line bundle on Δ_I.*

Proof. By lemma 1.5.20 and lemma 1.5.22 all fibers of $(\wedge^p \iota_I^* \mathcal{I}_I)^{\mathfrak{S}_I}$ are of rank one. This implies the assertion (see e.g. [Har77, II Ex. 5.8]). $\square$

Lemma 1.5.24. *For every $I \subset [n]$ with $\#I = k$ the set of left cosets is given by*

$$\mathfrak{S}_n / (\mathfrak{S}_I \times \overline{\mathfrak{S}_I}) = \{\mathfrak{S}_{I \to J} \mid J \subset [n], \#J = k\}$$

where

$$\mathfrak{S}_{I \to J} = \{\sigma \in \mathfrak{S}_n \mid \sigma(I) = J\}.$$

The right cosets are exactly the $\mathfrak{S}_{J \to I}$ with $\#J = k$.

Proof. The sets $\mathfrak{S}_{I \to J}$ are invariant under multiplication by $\mathfrak{S}_I \times \overline{\mathfrak{S}_I}$ on the right and consist of $k!(n-k)! = \#(\mathfrak{S}_I \times \overline{\mathfrak{S}_I})$ elements. Thus they are indeed left cosets. They are disjoint for distinct J and J'. The number of $J \subset [n]$ with $\#J = k$ is $\binom{n}{k}$. Because of

$$\binom{n}{k} \cdot k!(n-k)! = n! = \#\mathfrak{S}_n$$

all left cosets are of this form. The proof for the right cosets is analogous. $\square$

Lemma 1.5.25. *Let X be a smooth surface over $\mathbb{C}$. For $k - 2 \geq p \geq 0$ let*

$$t : L^{-2p} \iota_{[k]}^* \mathcal{O}_{\Delta_{[k-1]}} \cong \wedge^{2p} \iota_{[k]}^* \mathcal{I}_{[k-1]} \to \wedge^{2p} \iota_{[k]}^* \mathcal{I}_{[k]} \cong L^{-2p} \iota_{[k]}^* \mathcal{O}_{\Delta_{[k]}}$$

be the morphism induced by the natural surjection $\mathcal{O}_{\Delta_{[k-1]}} \to \mathcal{O}_{\Delta_{[k]}}$. We set $\sigma_k = \mathrm{id}_{[k]}$ and

$\sigma_\ell = (\ell,\, \ell+1, \ldots, k)$ *for* $\ell = 1, \ldots, k-1$. *Then*

$$T\colon [\wedge^{2p}\iota_{[k]}^*\mathcal{I}_{[k-1]}]^{\mathfrak{S}_{[k-1]}} \to [\wedge^{2p}\iota_{[k]}^*\mathcal{I}_{[k]}]^{\mathfrak{S}_{[k]}}\,,\ v \mapsto \sum_{\ell=1}^{k} \sigma_\ell t(v)$$

is an isomorphism.

Proof. Since both sides are line bundles on $\Delta_{[k]}$ the morphism T is an isomorphism if and only if it is fiberwise. So let $P \in \Delta_{[k]}$. We have $\sigma_\ell([k-1]) = [k] \setminus \{\ell\}$ so by the last lemma the σ_ℓ form a system of representatives of $\mathfrak{S}_{[k]}/\mathfrak{S}_{[k-1]}$. Thus for an $\mathfrak{S}_{[k-1]}$-invariant v the element $\sum_{\ell=1}^{k} \sigma_\ell t(v)$ is $\mathfrak{S}_{[k]}$-invariant and T is indeed well-defined. Since both spaces are one-dimensional by 1.5.22, it suffices to show that $T(P)$ is non-zero. Note that by proposition 1.5.22 and lemma 1.5.16 the map $t(P)$ is given by the wedge product of the natural inclusion $\varrho_{[k-1]} \to \varrho_{[k]}$. Now that $T(P)\colon (\wedge 2p\varrho_{[k-1]})^{\mathfrak{S}_{[k-1]}} \to (\wedge^{2p}\varrho_{[k]})$ is indeed an isomorphism follows from [Sca09a, Lemma C.6. (4)]. We will give another proof here. The standard representations $\varrho_{[k-1]}$ and $\varrho_{[k]}$ have the canonical bases given by $\zeta^j = e_j - e_{j-1}$ for $j \leq k-2$ respectively $j \leq k-1$. We consider the bases of $\wedge^{2p}(\varrho_{[k-1]}^{\oplus 2})$ and $\wedge^{2p}(\varrho_{[k]}^{\oplus 2})$ induced by those bases. We show that T maps every non-zero vector v whose coefficients in this base are non-negative real numbers to a non-zero vector with the same property. Since $t(P)$ is just the inclusion, $t(P)(v)$ is still non-zero with non-negative real coefficients. The action of σ_ℓ on the ζ^j is given by

$$\sigma_\ell \cdot \zeta^j = \begin{cases} \zeta^j & \text{if } j \leq \ell - 2 \\ \zeta^j + \zeta^{j+1} & \text{if } j = \ell - 1 \\ \zeta^{j+1} & \text{if } j \geq \ell. \end{cases}$$

Hence all the $\sigma_\ell t(P)(v)$ and thus also their sum $T(P)(v)$ have the desired property. For $k - 2 = p$ the invariants of $\wedge^{2p}(\varrho_{[k-1]}^{\oplus 2})$ are the whole space since it is one-dimensional. Thus, there is clearly an invariant non-zero vector with non-negative real coefficients. Now it follows by induction over k that $T(P)$ is indeed non-zero. $\qquad\square$

1.5.5 Normal varieties

In this subsection let M be a normal variety. For any open subvariety $\iota\colon U \hookrightarrow M$ with $\mathrm{codim}(M \setminus U, M) \geq 2$ and $F \in \mathrm{Coh}(X)$ a locally free sheaf $u_F\colon F \xrightarrow{\cong} \iota_*\iota^* F$ is an isomorphism. Here u_F is the unit of the adjunction (ι^*, ι_*), i.e. the morphism given by restriction of the sections.

Lemma 1.5.26. *Let N be another normal variety and $f\colon N \to M$ be a morphism such that also $\mathrm{codim}(N \setminus f^{-1}(U), N) \geq 2$ holds. Then there is a natural isomorphism*

$$\iota_*\iota^* f_* E \cong f_* E$$

38

for every locally free sheaf E on N.

Proof. Due to the flat base change

$$\begin{array}{ccc} f^{-1}(U) & \xrightarrow{\ \iota'\ } & N \\ {\scriptstyle f'}\downarrow & & \downarrow{\scriptstyle f} \\ U & \xrightarrow{\ \iota\ } & M \end{array}$$

we get indeed

$$\iota_* \iota^* f_* E \cong \iota_* f'_* \iota'^* E \cong f_* \iota'_* \iota'^* E \cong f_* E\,.$$

$\square$

Lemma 1.5.27. *Let $\iota\colon U \to X$ be any open immersion and let*

$$0 \to F' \to F \to F''$$

be an exact sequence in $\mathrm{Coh}(X)$ such that u_F and $u_{F''}$ are isomorphisms. Then $u_{F'}$ is also an isomorphism.

Proof. The functor $\iota^* \iota_*$ is left-exact. Therefore, there is the following diagram with exact horizontal sequences:

$$\begin{array}{ccccccc} 0 & \longrightarrow & F' & \longrightarrow & F & \longrightarrow & F'' \\ & & \downarrow & & {\scriptstyle\cong}\downarrow & & {\scriptstyle\cong}\downarrow \\ 0 & \longrightarrow & \iota_* \iota^* F' & \longrightarrow & \iota_* \iota^* F & \longrightarrow & \iota_* \iota^* F''\,. \end{array}$$

Since the last two vertical maps are isomorphisms, the first one is also. $\square$

Lemma 1.5.28. *Let X be a normal variety and $U \subset M$ an open subvariety such that $\mathrm{codim}(X \setminus U, X) \geq 2$. Given two locally free sheaves F and G on X and a subsheaf $E \subset F$ with $E_{|U} = F_{|U}$ the maps $a\colon \mathrm{Hom}(F,G) \to \mathrm{Hom}(E,G)$ and $\hat{a}\colon \mathcal{H}om(F,G) \to \mathcal{H}om(E,G)$, given by restricting the domain of the morphisms, are isomorphism.*

Proof. It suffices to show that a is an isomorphism. Since every open $V \subset X$ is again a normal variety with $V \setminus (U \cap V)$ of codimension at least 2, it follows by considering an open affine covering that $\hat{a}$ is also an isomorphism. We construct the inverse $b\colon \mathrm{Hom}(E,G) \to \mathrm{Hom}(F,G)$ of a. For a morphism $\varphi\colon E \to G$ the morphism $b(\varphi)$ sends $s \in F(V)$ to the unique section in $G(V)$ which restricts to $\varphi(s_{|V\cap U}) \in G(V \cap U)$. $\square$

Chapter 2

Image of tautological sheaves under the Bridgeland-King-Reid equivalence

2.1 The Bridgeland–King–Reid equivalence

Let G be a finite group acting on a smooth quasi-projective variety M over $\mathbb{C}$. The quotient $\pi \colon M \to M/G$ always exists as a quasi-projective variety but is in general singular. The following was introduced by Nakamura (see [Nak01]) as a candidate for a resolution of the singularities of M/G with good properties.

Definition 2.1.1. The *G-Hilbert scheme* $\mathrm{GHilb}(M)$ parametrises G-clusters on M, i.e. it is the scheme representing the functor

$$T \mapsto \left\{ \begin{array}{c} Z \subset T \times X \text{ closed} \\ \text{G-invariant subscheme} \end{array} \middle| \begin{array}{c} Z \text{ is flat and finite over } T, \\ H^0(Z_t) \simeq \mathbb{C}^G \text{ as } G\text{-representations } \forall t \in T \end{array} \right\}.$$

The *Nakamura-G-Hilbert scheme* $\mathrm{Hilb}^G(M)$ is defined to be the irreducible component of $\mathrm{GHilb}(M)$ which contains all the $\mathbb{C}$-valued points corresponding to free orbits. It is equipped with a universal family $\mathcal{Z} \subset \mathrm{Hilb}^G(M) \times M$ and the *G-Hilbert-Chow morphism*

$$\tau \colon \mathrm{Hilb}^G(M) \to M/G$$

sending a G-cluster to the G-orbit supporting it. In summary there is the commutative diagram

$$\begin{array}{ccc} \mathcal{Z} & \xrightarrow{\;p\;} & X^n \\ q\downarrow & & \downarrow\pi \\ \mathrm{Hilb}^G(M) & \xrightarrow{\;\tau\;} & S^nX \end{array}.$$

Theorem 2.1.2 ([BKR01]). *Let the quotient M/G be Gorenstein, i.e. $\omega_{M/G}$ exists as a line bundle, and*

$$\dim(\mathrm{Hilb}^G(M) \times_{M/G} \mathrm{Hilb}^G(M)) \le \dim(M) + 1\,.$$

Then τ is a crepant resolution of the singularities of M/G, i.e. $\tau^*\omega_{M/G} \cong \omega_{\mathrm{Hilb}^G(M)}$, and the equivariant Fourier-Mukai transform

$$\Phi_{\mathcal{O}_{\mathcal{Z}}} = Rp_* \circ q^*\colon\; \mathrm{D}^b(\mathrm{Hilb}^G(M)) \to \mathrm{D}^b_G(M)$$

is an equivalence of triangulated categories.

The functor $\Phi_{\mathcal{O}_{\mathcal{Z}}}$ is called *the Bridgeland-King-Reid equivalence*.

Remark 2.1.3. The group G acts trivially on $\mathrm{Hilb}^G(M)$ and on $\mathcal{Z} \subset \mathrm{Hilb}^G(M) \times M$ by the action induced by the action on M. To be formally correct the equivariant Fourier-Mukai transform $\Phi_{\mathcal{O}_{\mathcal{Z}}}$ is the functor $Rp_* \circ q^*\colon \mathrm{D}^b_G(\mathrm{Hilb}^G(M)) \to \mathrm{D}^b_G(M)$ and the Bridgeland-King-Reid equivalence is the functor $\Phi_{\mathcal{O}_{\mathcal{Z}}} \circ \mathrm{triv}$.

2.2 The Hilbert scheme of points on a surface

Let X be a quasi-projective surface over $\mathbb{C}$ and $P \in \mathbb{Q}[x]$. Then the *Hilbert functor*

$$T \mapsto \left\{ \begin{array}{c|c} Z \subset T \times X & Z \text{ is flat and proper over } T, \\ \text{closed subscheme} & Z_t \text{ has Hilbert polynomial } P \;\forall t \in T \end{array} \right\}$$

is representable by a result of Grothendieck (see [Gro95]). In case $P = n \in \mathbb{N}$ the representing scheme $X^{[n]}$ is called the *Hilbert scheme of n points on X*. Its $\mathbb{C}$-valued points correspond to zero-dimensonal subschemes ξ of X of length

$$\ell(\xi) := \dim_{\mathbb{C}} \Gamma(\xi, \mathcal{O}_\xi) = n\,.$$

The generic case of such subschemes are collections of n distinct points. If the support of ξ consists of less than n points, the scheme structure has to be further specified. For example a subscheme of length 2 which is concentrated in one point is given by choosing a tangent direction of X in that point. $X^{[n]}$ is a $2n$-dimensional quasi-projective smooth variety (see

41

[Fog68]). It is a resolution of the symmetric product $S^n X := X^n/\mathfrak{S}_n$ via the *Hilbert-Chow morphism*

$$\mu \colon X^{[n]} \to S^n X \quad, \quad \xi \mapsto \sum_{x \in \xi} \ell(\xi, x) \cdot x \,.$$

The *isospectral Hilbert scheme* is defined as $I^n X := (X^{[n]} \times_{S^n X} X^n)_{\mathrm{red}}$. It is a family of $\mathfrak{S}_n$-clusters in X^n flat over $X^{[n]}$ (see [Hai01]). Thus it induces a morphism η from $X^{[n]}$ to the moduli space $\mathfrak{S}_n \operatorname{Hilb}(X^n)$. Since the generic fiber of $I^n X$ over $X^{[n]}$ consist of n distinct points, η factorises over $\operatorname{Hilb}^{\mathfrak{S}_n}(X^n)$.

Theorem 2.2.1 ([Hai01]). *The morphism η induces an isomorphism between the commutative diagrams*

$$
\begin{array}{ccc}
I^n X & \xrightarrow{\ p\ } & X^n \\
{\scriptstyle q}\downarrow & & \downarrow{\scriptstyle \pi} \\
X^{[n]} & \xrightarrow[\ \mu\ ]{} & S^n X
\end{array}
\qquad \cong \qquad
\begin{array}{ccc}
\mathcal{Z} & \xrightarrow{\ p\ } & X^n \\
{\scriptstyle q}\downarrow & & \downarrow{\scriptstyle \pi} \\
\operatorname{Hilb}^{\mathfrak{S}_n}(X^n) & \xrightarrow[\ \tau\ ]{} & S^n X
\end{array} \,.
$$

That means that $X^{[n]}$ with $I^n X$ in the role of the universal family of $\mathfrak{S}_n$-clusters is isomorphic to $\operatorname{Hilb}^{\mathfrak{S}_n}(X^n)$ over $S^n X$. Since $S^n X$ is Gorenstein and the Hilbert-Chow morphism is semi-small (see also [Hai01]) the assumptions of the Bridgeland-King-Reid theorem are fulfilled.

Corollary 2.2.2. *The Fourier-Mukai Transform*

$$\Phi := \Phi^{X^{[n]} \to X^n}_{\mathcal{O}_{I^n X}} \colon \operatorname{D}^b(X^{[n]}) \to \operatorname{D}^b_{\mathfrak{S}_n}(X^n)$$

is an equivalence of triangulated categories.

The Bridgeland–King–Reid equivalence can be used to compute the extension groups of objects in the derived category of the Hilbert scheme.

Corollary 2.2.3. *Let $\mathcal{F}^\bullet, \mathcal{G}^\bullet \in \operatorname{D}^b(X^{[n]})$. Then*

$$\operatorname{Ext}^i_{X^{[n]}}(\mathcal{F}^\bullet, \mathcal{G}^\bullet) \cong \mathfrak{S}_n \operatorname{Ext}^i_{X^n}(\Phi(\mathcal{F}^\bullet), \Phi(\mathcal{G}^\bullet)) \quad \text{for all } i \in \mathbb{Z} \,.$$

Proof. Using the last corollary we indeed have

$$
\begin{aligned}
\operatorname{Ext}^i_{X^{[n]}}(\mathcal{F}^\bullet, \mathcal{G}^\bullet) &\cong \operatorname{Hom}_{\operatorname{D}^b(X^{[n]})}(\mathcal{F}^\bullet, \mathcal{G}^\bullet[i]) \cong \operatorname{Hom}_{\operatorname{D}^b_{\mathfrak{S}_n}(X^n)}(\Phi(\mathcal{F}^\bullet), \Phi(\mathcal{G}^\bullet[i])) \\
&\cong \operatorname{Hom}_{\operatorname{D}^b_{\mathfrak{S}_n}(X^n)}(\Phi(\mathcal{F}^\bullet), \Phi(\mathcal{G}^\bullet)[i]) \\
&\cong \mathfrak{S}_n \operatorname{Ext}^i_{X^n}(\Phi(\mathcal{F}^\bullet), \Phi(\mathcal{G}^\bullet)) \,.
\end{aligned}
$$

$\square$

We will abbreviate the functor $[\_]^{\mathfrak{S}_n} \circ \pi_* \colon D^b_{\mathfrak{S}_n}(X^n) \to D^b(S^n X)$ by $[\_]^{\mathfrak{S}_n}$. Note that π_* indeed does not need to be derived since π is finite.

Proposition 2.2.4 ([Sca09a]). *For every $\mathcal{F}^\bullet \in D^b(X^{[n]})$ there is a natural isomorphism*

$$R\mu_* \mathcal{F}^\bullet \simeq [\Phi(\mathcal{F}^\bullet)]^{\mathfrak{S}_n}.$$

Furthermore there are natural isomorphisms

$$\mathrm{H}^*(X^{[n]}, \mathcal{F}^\bullet) \cong \mathrm{H}^*(S^n X, \Phi(\mathcal{F}^\bullet)^{\mathfrak{S}_n}) \cong \mathrm{H}^*(X^n, \Phi(F^\bullet))^{\mathfrak{S}_n}.$$

Proof. The morphisms q is the $\mathfrak{S}_n$-quotient morphisms (see [Sca09a, section 1.5]). Thus, $(q_* \mathcal{O}_{I^n X})^{\mathfrak{S}_n} = \mathcal{O}_{X^{[n]}}$. Using this, we get

$$
\begin{aligned}
\Phi(\mathcal{F}^\bullet)^{\mathfrak{S}_n} &\simeq [\pi_* \circ Rp_* \circ q^* \mathcal{F}^\bullet]^{\mathfrak{S}_n} \simeq [R\mu_* \circ q_* \circ q^* \mathcal{F}^\bullet]^{\mathfrak{S}_n} \simeq R\mu_* [q_* \circ q^* \mathcal{F}^\bullet]^{\mathfrak{S}_n} \\
&\overset{PF}{\simeq} R\mu_* [\mathcal{F}^\bullet \otimes^L q_* \mathcal{O}_{I^n X}]^{\mathfrak{S}_n} \\
&\overset{1.5.9}{\simeq} R\mu_* (\mathcal{F}^\bullet \otimes^L [q_* \mathcal{O}_{I^n X}]^{\mathfrak{S}_n}) \\
&\simeq R\mu_* \mathcal{F}^\bullet.
\end{aligned}
$$

Now indeed

$$\mathrm{H}^*(X^{[n]}, \mathcal{F}^\bullet) \cong \mathrm{H}^*(S^n X, R\mu_* \mathcal{F}^\bullet) \cong \mathrm{H}^*(S^n X, \Phi(\mathcal{F}^\bullet)^{\mathfrak{S}_n}).$$

The last isomorphisms of the second assertion is due to the fact that $[\_]^{\mathfrak{S}_n}$ is an exact functor. $\qquad\square$

2.3 Tautological sheaves

Definition 2.3.1. We define the *tautological functor for sheaves* as

$$(\_)^{[n]} := \mathrm{pr}_{X^{[n]}*}(\mathcal{O}_\Xi \otimes \mathrm{pr}_X^*(\_)) \colon \mathrm{Coh}(X) \to \mathrm{Coh}(X^{[n]}).$$

For a sheaf $F \in \mathrm{Coh}(X)$ we call its image $F^{[n]}$ under this functor the *tautological sheaf associated with F*. In [Sca09b, Proposition 2.3] it is shown that the functor $(\_)^{[n]}$ is exact. Thus it induces the *tautological functor for objects* $(\_)^{[n]} \colon D^b(X) \to D^b(X^{[n]})$. For an object $F^\bullet \in D^b(X)$ the *tautological object associated to $F^\bullet$* is $(F^\bullet)^{[n]}$.

Remark 2.3.2. The tautological functor for objects is isomorphic to the Fourier-Mukai transform with kernel the structural sheaf of the universal family, i.e. $(F^\bullet)^{[n]} \simeq \Phi^{X \to X^{[n]}}_{\mathcal{O}_\Xi}(F^\bullet)$ for every $F^\bullet \in D^b(X)$.

Remark 2.3.3. If F is locally free of rank k the *tautological bundle* $F^{[n]}$ is locally free of rank $k \cdot n$ with fibers $F^{[n]}([\xi]) = \Gamma(\xi, F_{|\xi})$ since $\mathrm{pr}_{X^{[n]}} \colon \Xi \to X^{[n]}$ is flat and finite of degree n.

2.4 The complex $C^\bullet$

We use the notation from section 1.2. To any coherent sheaf F on X we associate a $\mathfrak{S}_n$-equivariant complex $C_F^\bullet$ of sheaves on X^n as follows. Remember that $F_I = \iota_{I*} p_I^* F$. We set

$$C_F^0 = \bigoplus_{i=1}^n p_i^* F \quad , \quad C_F^p = \bigoplus_{I \subset [n], |I| = p+1} F_I \quad \text{for } 0 < p < n \quad , \quad C_F^p = 0 \quad \text{else}.$$

Let $s = (s_I)_{|I|=p+1}$ be a local section of C_F^p. We define the $\mathfrak{S}_n$-linearization of C_F^p by

$$\lambda_\sigma(s)_I := \varepsilon_{\sigma,I} \cdot \sigma_*(s_{\sigma^{-1}(I)}),$$

where σ_* is the flat base change isomorphism from the following diagram with $p_I \circ \sigma = p_{\sigma^{-1}(I)}$

$$
\begin{array}{ccc}
\Delta_{\sigma^{-1}(I)} & \xrightarrow{\sigma} \Delta_I \xrightarrow{p_I} & X \\
{\scriptstyle \iota_{\sigma^{-1}(I)}} \downarrow & \downarrow {\scriptstyle \iota_I} & \\
X^n & \xrightarrow{\sigma} X^n . &
\end{array}
$$

This gives also a $\mathfrak{S}_n$-linearization of C_F^0 using the convention $F_{\{i\}} := p_i^* F$ and $\Delta_{\{i\}} := X^n$. Finally, we define the differentials $d^p \colon C_F^p \to C_F^{p+1}$ by the formula

$$d^p(s)_J := \sum_{i \in J} \varepsilon_{i,J} \cdot s_{J \setminus \{i\}|_{\Delta_J}}.$$

As one can check using lemma 1.2.1, $C_F^\bullet$ is indeed an $\mathfrak{S}_n$-equivariant complex.

Remark 2.4.1. Since p_I is a projection and ι_I a closed embedding, the functor

$$C^p \colon \mathrm{Coh}(X) \to \mathrm{Coh}_{\mathfrak{S}_n}(X^n) \quad , \quad F \mapsto C_F^p$$

is exact for all $0 \le p \le n-1$. Thus, the functor

$$C^\bullet \colon \mathrm{Kom}(\mathrm{Coh}(X)) \to \mathrm{Kom}(\mathrm{Coh}_{\mathfrak{S}_n}(X^n)) \quad F^\bullet \mapsto C_{F^\bullet}^\bullet := \mathrm{tot}(K^{\bullet,\bullet})$$

is also exact, where the double complex is given by $K^{i,j} = C_{F^j}^i$. Hence, without deriving we get an exact functor $C^\bullet \colon \mathrm{D}^b(X) \to \mathrm{D}_{\mathfrak{S}_n}^b(X^n)$.

Remark 2.4.2. For every $I \subset [n]$ the sheaf F_I carries a $\mathrm{Stab}(I) = \mathfrak{S}_I \times \overline{\mathfrak{S}}_I$-linearization. As

44

a $\mathfrak{S}_I \times \overline{\mathfrak{S}_I}$-sheaf we can write it with the notation of subsection 1.4.8 as

$$F_I = \iota_{I*}(\mathfrak{a} \otimes p_I^* F).$$

See also remark 1.4.4 for a description of the linearization of $p_I^* F$. Note that, since the group $\mathfrak{S}_I$ acts trivially on Δ_I, the $\mathfrak{S}_I$-linearization of F_I is just a $\mathfrak{S}_I$-action. An element $\sigma \in \mathfrak{S}_I$ acts by multiplication by $\mathrm{sgn}(\sigma)$ on F_I.

Remark 2.4.3. For $1 \leq \ell \leq n$ we fix an $I_0 \subset [n]$ with ℓ elements, e.g. $I_0 = [\ell]$, and choose for any $I \subset [n]$ with $\#I = \ell$ a $\sigma \in \mathfrak{S}_n$ with $\sigma(I) = I_0$, e.g. $\sigma = {}_{I_0}u_I \times {}_{\overline{I}_0}u_{\overline{I}}$. Then the chosen σ form a system of representatives of $(\mathfrak{S}_{I_0} \times \overline{\mathfrak{S}_{I_0}}) \setminus \mathfrak{S}_n$ (lemma 1.5.24) and the canonical isomorphisms $F_{\Delta_I} \cong \sigma^* F_{\Delta_{I_0}}$ induce an isomorphism (see also lemma 1.4.1)

$$C_F^{\ell-1} \cong \mathrm{Inf}_{\mathfrak{S}_{I_0} \times \overline{\mathfrak{S}_{I_0}}}^{\mathfrak{S}_n}(F_{I_0}).$$

2.5 Polygraphs and the image of tautological sheaves under Φ

Let $E_1^\bullet, \ldots, E_k^\bullet \in \mathrm{D}^b(X)$. We define $\Xi(n,k)$ as the k-fold fiber product

$$\Xi(n,k) := \Xi \times_{X^{[n]}} \cdots \times_{X^{[n]}} \Xi$$

which is a closed subscheme of $X^{[n]} \times X^k$. By base change in the cartesian diagram

$$
\begin{array}{ccccc}
\Xi(n,k) & \longrightarrow & \Xi^k & \xrightarrow{\mathrm{pr}_X^k} & X^k \\
\downarrow & & {\scriptstyle \mathrm{pr}_{X^{[n]}}^k}\downarrow & & \\
X^{[n]} & \xrightarrow{\ i\ } & (X^{[n]})^k & &
\end{array}
$$

where the lower horizontal arrow i is the diagonal embedding we get natural isomorphisms

$$(E_1^\bullet)^{[n]} \otimes^L \cdots \otimes^L (E_k^\bullet)^{[n]} \simeq Li^* \mathrm{pr}_{X^{[n]}*}^k \mathrm{pr}_X^{k*}(E_1^\bullet \boxtimes \cdots \boxtimes E_k^\bullet)$$
$$\simeq \Phi_{\mathcal{O}_{\Xi(n,k)}}^{X^k \to X^{[n]}}(E_1^\bullet \boxtimes \cdots \boxtimes E_k^\bullet).$$

Thus, the image of the tensor product of tautological objects under the Bridgeland–King–Reid equivalence is given by

$$\Phi(E_1^\bullet \otimes^L \cdots \otimes^L E_k^\bullet) \simeq \Phi_{\mathcal{O}_{I^n X}}^{X^{[n]} \to X^n} \circ \Phi_{\mathcal{O}_{\Xi(n,k)}}^{X^k \to X^{[n]}}(E_1^\bullet \boxtimes \cdots \boxtimes E_k^\bullet)$$
$$\simeq \Phi_{\mathcal{P}}^{X^k \to X^n}(E_1^\bullet \boxtimes \cdots \boxtimes E_k^\bullet).$$

where $\mathcal{P} = Rp_{13*}(p_{12}^*\mathcal{O}_{\Xi(n,k)} \otimes^L p_{23}^*\mathcal{O}_{I^nX})$ is the Fourier-Mukai kernel of the composition. Here p_{ij} denotes the projection form $X^k \times X^{[n]} \times X^n$ to the product of the i-th and the j-th factor. Since $\Xi(n,k)$ is flat over $X^{[n]}$ the tensor product occurring in $\mathcal{P}$ needs not to be derived and $p_{12}^*\mathcal{O}_{\Xi(n,k)} \otimes^L p_{23}^*\mathcal{O}_{I^nX} \cong \mathcal{O}_{Z(n,k)}$ where $Z(n,k) := \Xi(n,k) \times_{X^{[n]}} I^nX$. We define the reduced closed subscheme $D = D(n,1)$ of $X^n \times X$ by

$$D := \{(x_1,\ldots,x_n,y) \mid y \in \{x_1,\ldots,x_n\}\}.$$

Furthermore, we define $\widetilde{D(n,k)}$ as the k-fold fiber product

$$\widetilde{D(n,k)} := D \times_{X^n} \cdots \times_{X^n} D.$$

It is a closed subscheme of $X^n \times X^k$ which is non-reduced for $k \geq 2$. We define the *polygraph* by $D(n,k) := \widetilde{D(n,k)}_{\mathrm{red}}$. The $\mathbb{C}$-valued points of the polygraph are given by

$$D(n,k) = \{(x_1,\ldots,x_n,y_1,\ldots,y_k) \mid \{y_1,\ldots,y_k\} \subset \{x_1,\ldots,x_n\}\}.$$

Haiman's vanishing theorem states that $Rp_{13*}\mathcal{O}_{Z(n,k)} \cong \mathcal{O}_{D(n,k)}$ (see [Hai02] for the case $X = \mathbb{A}^2$ and [Sca09a, Theorem 1.6.2] for the generalisation to X an arbitrary smooth quasi-projective surface). In summary, we get the following.

Proposition 2.5.1 ([Sca09b]). *The image of the derived tensor product of tautological objects under the Bridgeland–King–Reid equivalence is given by the natural isomorphism*

$$\Phi((E_1^\bullet)^{[n]} \otimes^L \cdots \otimes^L (E_k^\bullet)^{[n]}) \simeq \Phi_{\mathcal{O}_{D(n,k)}}^{X^k \to X^n}(E_1^\bullet \boxtimes \cdots \boxtimes E_k^\bullet).$$

Corollary 2.5.2. *Let $E_1,\ldots,E_k$ be locally free sheaves on X. Then the image of the tensor product of the induced tautological bundles under the Bridgeland–King–Reid equivalence is cohomologically concentrated in degree zero, i.e.*

$$\Phi(E_1^{[n]} \otimes \cdots \otimes E_k^{[n]}) \simeq p_*q^*(E_1^{[n]} \otimes \cdots \otimes E_k^{[n]}).$$

Thus, also $R\mu_*(E_1^{[n]} \otimes \cdots \otimes E_k^{[n]}) \simeq \mu_*(E_1^{[n]} \otimes \cdots \otimes E_k^{[n]})$.

Proof. Since the tautological bundles $E_i^{[n]}$ are locally free sheaves on $X^{[n]}$ we have

$$E_1^{[n]} \otimes^L \cdots \otimes^L E_k^{[n]} \simeq E_1^{[n]} \otimes \cdots \otimes E_k^{[n]}.$$

Thus, by the previous proposition $\Phi(E_1^{[n]} \otimes \cdots \otimes E_k^{[n]})$ is given by

$$\Phi_{\mathcal{O}_{D(n,k)}}(E_1 \boxtimes \cdots \boxtimes E_k) = Rp_{X^n*}(\mathcal{O}_{D(n,k)} \otimes^L p_{X^k}^*(E_1 \boxtimes \cdots \boxtimes E_k)).$$

Since the E_i are locally free, the tensor product needs not to be derived. Since $D(n, k)$ is finite over X^n the push-forward needs not to be derived. Since $\Phi \simeq Rp_* \circ q^*$ indeed

$$\Phi(E_1^{[n]} \otimes \cdots \otimes E_k^{[n]}) \simeq p_* q^*(E_1^{[n]} \otimes \cdots \otimes E_k^{[n]}).$$

The last assertion follows by proposition 2.2.4 and the fact that the functor $[\_]^{\mathfrak{S}_n}$ is exact. $\square$

See [Sca09b] for weaker conditions on the sheaves $E_i \in \mathrm{Coh}(X)$ under which the tensor product $E_1^{[n]} \otimes^L \cdots \otimes^L E_k^{[n]}$ respectively its image under Φ are cohomologically concentrated in degree zero. The irreducible components of the Polygraph $D = D(n, 1)$ are given by

$$D_i = \{(x_1, \ldots, x_n, y) \in X^n \times X \mid y = x_i\} \subset X^n \times X$$

for $i = 1, \ldots, n$. For $I \subset [n]$ we set $D_I = \cap_{i \in I} D_i$. Then a right resolution $\mathcal{K}^\bullet$ of $\mathcal{O}_D$ is given by $\mathcal{K}^p = \oplus_{\#I=p+1} \mathcal{O}_{D_I}$ with the differentials given by the restrictions $\mathcal{O}_{D_I} \to \mathcal{O}_{D_{I \cup \{i\}}}$ together with signs as in the complex $C_F^\bullet$. The augmentation map $\hat{\gamma}: \mathcal{O}_D \to \oplus_{i=1}^n \mathcal{O}_{D_i}$ is also given by restriction of the sections to the irreducible components. The reason for $\mathcal{K}^\bullet$ being indeed a resolution of $\mathcal{O}_D$ is that the components D_i are smooth and hence Cohen-Macauley and that they intersect properly (see [Sca09a, Appendix A]).

Theorem 2.5.3 ([Sca09b]). *For every* $F \in \mathrm{Coh}(F)$ *the object* $\Phi(F^{[n]})$ *is cohomologically concentrated in degree zero. Furthermore, the complex* $C_F^\bullet$ *is a right resolution of* $p_* q^*(F^{[n]})$. *Hence, in* $\mathrm{D}_{\mathfrak{S}_n}^b(X^n)$ *there are the isomorphisms*

$$\Phi(F^{[n]}) \simeq p_* q^* F^{[n]} \simeq C_F^\bullet.$$

Proof. For $I \subset [n]$ we have over X the isomorphism $D_I \cong X \times X^{n-|I|}$. Thus, all the terms $\mathcal{K}^p$ for $p \geq 0$ are flat over X. Since $\mathcal{K}^\bullet$ is a right resolution of $\mathcal{O}_D$, it follows that $\mathcal{O}_D$ is also flat over X. Furthermore, D is finite over X^n. Thus, $\Phi(F^{[n]}) \simeq \Phi_{\mathcal{O}_D}^{X \to X^n}(F)$ is indeed concentrated in degree 0 and can be computed as

$$\Phi(F^{[n]}) \simeq \Phi_{\mathcal{O}_D}^{X \to X^n}(F) \simeq p_{X^n *}(p_X^* F \otimes \mathcal{K}^\bullet).$$

Now $p_{X^n|D_I}: D_I \to \Delta_I$ is an isomorphism which yields $p_{X^n *}(p_X^* F \otimes \mathcal{O}_{D_I}) \cong F_I$ and hence the second isomorphism of the theorem. $\square$

Remark 2.5.4. Every morphism $\varphi: E \to F$ of coherent sheaves on X induces a morphism $\varphi^{[n]} = \mathrm{pr}_{X^{[n]}*}(\mathcal{O}_\Xi \otimes \mathrm{pr}_X^* \varphi): E^{[n]} \to F^{[n]}$. Under the isomorphism of the theorem

$$p_* q^*(\varphi^{[n]}) \simeq \Phi(\varphi^{[n]}) \simeq \Phi_{\mathcal{O}_D}^{X \to X^n}(\varphi) \simeq p_{X^n *}(p_X^* \varphi \otimes \mathcal{K}^\bullet).$$

corresponds to $C^\bullet(\varphi)\colon C^\bullet_E \to C^\bullet_F$. Thus we can rephrase the theorem by saying that there is an isomorphism of functors

$$p_* q^*(\_)^{[n]} \cong C^\bullet\colon \operatorname{Coh}(X) \to \operatorname{Kom}(\operatorname{Coh}_{\mathfrak{S}_n}(X^n))\,.$$

The functors $(\_)^{[n]}$ and q^* are exact (q is flat). Furthermore, $q^* F^{[n]}$ is p_*-acyclic for every tautological sheaf $F^{[n]}$ by the above theorem. Thus for every $E^\bullet \in D^b(X)$ we have a natural isomorphism $\Phi((E^\bullet)^{[n]}) \simeq p_* q^*(E^{[n]})^\bullet$. Since $C^\bullet$ is also an exact functor, we have the following isomorphism of functors on the level of derived categories:

$$\Phi((\_)^{[n]}) \simeq C^\bullet\colon D^b(X) \to D^b_{\mathfrak{S}_n}(X^n)\,.$$

So $\Phi((F^\bullet)^{[n]}) \simeq C^\bullet_{F^\bullet}$ holds for every $F^\bullet \in D^b(X)$.

Theorem 2.5.5 ([Sca09b])**.** *The cohomology of tautological sheaves is given by the following isomorphism of graded vector spaces*

$$H^*(X^{[n]}, F^{[n]}) \simeq H^*(F) \otimes S^{n-1}(H^*(\mathcal{O}_X))\,.$$

Proof. The sheaves C^p_F do not have any $\mathfrak{S}_n$-invariant sections for $p \geq 1$. The reason is that $\mathfrak{S}_I$ acts alternating on F_I for every $I \subset [n]$ with $|I| \geq 2$ (see remark 2.4.2). Thus, in $D^b(S^n X)$ we have $[\pi_* \Phi(F^{[n]})]^{\mathfrak{S}_n} \simeq [\pi_* C^0_F]^{\mathfrak{S}_n}$. Hence, with 2.2.4 it follows that

$$\mathrm{H}^*(X^{[n]}, F^{[n]}) \simeq [\mathrm{H}^*(X^n, C^0_F)]^{\mathfrak{S}_n}\,.$$

The invariants of $H^*(X^n, C^0_F)$ are computed using Danila's lemma (see subsection 1.5.2). Similar computations will be done in detail later for example in the proof of theorem 4.3.1. $\square$

2.6 Description for $k \geq 2$

Let $E_1, \ldots, E_k$ be locally free sheaves on X. We consider the cartesian diagram

$$
\begin{array}{ccc}
\widetilde{D(n,k)} & \longrightarrow & D^k \xrightarrow{\ \operatorname{pr}^k_X\ } X^k \\
\downarrow & & \operatorname{pr}^k_{X^n} \downarrow \\
X^n & \xrightarrow{\ i\ } & (X^n)^k
\end{array}
$$

where i denotes the diagonal embedding. We have a natural isomorphism

$$\Phi(E_1^{[n]}) \otimes^L \cdots \otimes^L \Phi(E_1^{[n]}) \simeq Li^* \operatorname{pr}^k_{X^n *} \operatorname{pr}^{k*}_X(E_1 \boxtimes \cdots \boxtimes E_k)\,.$$

The base change diagram above yields a morphism

$$B\colon Li^* \operatorname{pr}^k_{X^n*} \operatorname{pr}^{k*}_X(E_1 \boxtimes \cdots \boxtimes E_k) \to \Phi^{X^k \to X^n}_{\mathcal{O}_{\widetilde{D(n,k)}}}(E_1 \boxtimes \cdots \boxtimes E_k).$$

Furthermore, the natural surjection $\mathcal{O}_{\widetilde{D(n,k)}} \to \mathcal{O}_{D(n,k)}$ of the Fourier-Mukai kernels induces a natural transformation

$$A\colon \Phi^{X^k \to X^n}_{\mathcal{O}_{\widetilde{D(n,k)}}} \to \Phi^{X^k \to X^n}_{\mathcal{O}_{D(n,k)}}.$$

In degree zero the composition $A(\boxtimes_i E_i) \circ B$ gives a surjection

$$\alpha\colon p_*q^*(E_1^{[n]}) \otimes \cdots \otimes p_*q^*(E_1^{[n]}) \to p_*q^*(E_1^{[n]} \otimes \cdots \otimes E_k^{[n]})$$

with the image being torsion-free (see [Sca09a, p. 12]). Since α is an isomorphism outside of the big diagonal $\mathbb{D}$, its kernel is supported on $\mathbb{D}$ and hence torsion. Thus, the kernel is exactly the torsion subsheaf of $\otimes_i p_*q^*(E_i^{[n]})$. We denote for $i \in [k]$ the augmentation map by $\gamma_i\colon p_*q^*(E_i^{[n]}) \to C^0_{E_i}$, i.e. $\gamma_i = \operatorname{pr}_{X^n}(\hat{\gamma} \otimes \operatorname{id}_{\operatorname{pr}^*_X E_i})$.

Proposition 2.6.1. *Let $E_1, \ldots, E_k$ be locally free sheaves on X. Then there is a natural isomorphism*

$$p_*q^*(E_1^{[n]} \otimes \cdots \otimes E_k^{[n]}) \cong \operatorname{im}(\gamma_1 \otimes \cdots \otimes \gamma_k) \subset C^0_{E_1} \otimes \cdots \otimes C^0_{E_k}.$$

Proof. The sheaf $\otimes_i C^0_{E_i}$ is locally free and hence torsion-free. Since $C^1_{E_i}$ is supported on $\mathbb{D}$ for every $i \in [k]$, the map $\otimes_i \gamma_i$ is an isomorphism outside of $\mathbb{D}$. Thus, also $\operatorname{im}(\otimes_i \gamma_i)$ is the quotient of $\otimes_i p_*q^*(E_i^{[n]})$ by the torsion subsheaf, which leads to the identification with $p_*q^*(\otimes_i E_i^{[n]})$. $\square$

Remark 2.6.2. The morphism $\otimes_i \gamma_i\colon \otimes_i p_*q^*(E_i^{[n]}) \to \otimes_i C^0_{E_i}$ is $\mathfrak{S}_n$-equivariant since all the γ_i are equivariant. The morphism B is equivariant since it is induced by a base change diagram where all the arrows are equivariant and A is equivariant since it is induced by the surjection $\mathcal{O}_{\widetilde{D(n,k)}} \to \mathcal{O}_{D(n,k)}$ of equivariant Fourier-Mukai kernels. Thus, also α is equivariant. Hence, the isomorphism of the previous proposition is $\mathfrak{S}_n$-equivariant when considering $\operatorname{im}(\otimes_i \gamma_i)$ with the linearization induced by the linearization of $\otimes_i C^0_{E_i}$. By similar reasons, the isomorphism is also $\mathfrak{S}_k$-equivariant in the case $E_1 = \cdots = E_k = E$ when considering $\otimes_i E_i^{[n]}$ as well as $\otimes_i C^0_{E_i}$ with the action given by permuting the tensor factors.

2.7 Multitor spectral sequence

In [Sca09a] respectively [Sca09b] Scala gave another description of $p_*q^*(\otimes_i E_i^{[n]})$ as a subsheaf of $\otimes_i C^0_{E_i}$, namely as the $E^{0,0}_\infty$ term of a certain spectral sequence. Though we will not use this spectral sequence in the following proofs, many ideas how the results should look like came from the study of this spectral sequence. Furthermore proposition 2.6.1 can also be obtained from Scala's result as we explain in the following.

Theorem 2.7.1 ([Sca09b]). *Let $F_1, \ldots, F_k$ be locally free sheaves on X and let E be the multitor spectral sequence associated to the complexes $C_{F_1}^{\bullet}, \ldots, C_{F_k}^{\bullet}$ on X^n. Then we have $p_* q^* (\otimes_{i=1}^k F_i^{[n]}) \cong E_{\infty}^{0,0}$.*

We will now construct and explain the multitor spectral sequence. See also the remarks 2.3.3 and 4.1.1. of [Sca09a]. Let $C_1^{\bullet}, \ldots, C_k^{\bullet} \in D^b(X)$ be bounded complexes on a scheme with enough locally free sheaves. Let the double complexes $R_i^{\bullet,\bullet}$ for $i = 1, \ldots, k$ be locally free resolutions of $C_i^{\bullet}$. That means that $R_i^{p,q} = 0$ for $q > 0$ and that there are morphisms of complexes $p_i \colon R_i^{\bullet,0} \to C_i^{\bullet}$ which make $R_i^{k,\bullet}$ a right resolution of C_i^k via p_i^k for every $k \in \mathbb{Z}$. We define the double complex $A^{\bullet,\bullet}$ by

$$A^{p,q} := \bigoplus_{\substack{i_1 + \cdots + i_k = p, \\ j_1 + \cdots + j_k = q}} R_1^{i_1, j_1} \otimes \cdots \otimes R_k^{i_k, j_k}$$

with the differentials induced by those of the $R_i^{\bullet,\bullet}$ together with appropriate signs (for details see [Sca09a, Remark 4.1.1]) such that $A^{\bullet,\bullet}$ is a double complex. Then for every $i \in \mathbb{Z}$ we have $A^{i,\bullet} = \oplus_{i_1 + \cdots + i_k = i} R_1^{i_1, \bullet} \otimes \cdots \otimes R_k^{i_k, \bullet}$ and moreover

$$\text{tot}(A^{\bullet,\bullet}) = \text{tot}(R_1^{\bullet,\bullet}) \otimes \cdots \otimes \text{tot}(R_k^{\bullet,\bullet}).$$

Furthermore, the morphisms p_i induce a morphism $A^{\bullet,0} \to C_1^{\bullet} \otimes \cdots \otimes C_k^{\bullet}$ which in turn induces the morphism

$$P \colon C_1^{\bullet} \otimes^L \cdots \otimes^L C_k^{\bullet} \simeq \text{tot}(A^{\bullet,\bullet}) \xrightarrow{P} C_1^{\bullet} \otimes \cdots \otimes C_k^{\bullet}.$$

which belongs to the multi-derived tensor product functor. We define the *multitor spectral sequence* associated to $C_1^{\bullet}, \ldots, C_k^{\bullet}$ as the spectral sequence associated to $A^{\bullet,\bullet}$ with 1-level

$$E_1^{p,q} = \mathcal{H}^q(A^{p,\bullet}) = \bigoplus_{i_1 + \cdots + i_k = p} \text{Tor}_{-q}(C_1^{i_1}, \ldots, C_k^{i_k})$$

and converging to

$$E^n = \mathcal{H}^n(\text{tot } A^{\bullet,\bullet}) = \mathcal{H}^n(C_1^{\bullet} \otimes^L \cdots \otimes^L C_k^{\bullet}).$$

For the next proposition we assume that all the $C_i^{\bullet}$ are right resolutions of sheaves F_i of the form

$$0 \to F_i \xrightarrow{\gamma_i} \to C_i^0 \xrightarrow{d_i^0} C_i^1 \to \ldots.$$

Then the map $\otimes_{i=1}^{k} p_i^0$ induces an isomorphism

$$E_1^{0,0} \xrightarrow{\cong} C_1^0 \otimes \cdots \otimes C_k^0.$$

Using this isomorphism we can consider canonically $E_\infty^{0,0} \subset E_1^{0,0}$ as a subsheaf of $C_1^0 \otimes \cdots \otimes C_k^0$.

Proposition 2.7.2. *The sheaf $E_\infty^{0,0}$ coincides as a subsheaf of $C_1^0 \otimes \cdots \otimes C_k^0$ with the image of the morphism*

$$\gamma_1 \otimes \cdots \otimes \gamma_k \colon F_1 \otimes \cdots \otimes F_k \to C_1^0 \otimes \cdots \otimes C_k^0.$$

Proof. By the consideration above, the degree zero part P^0 of the natural morphism

$$C_1^\bullet \otimes^L \cdots \otimes^L C_k^\bullet \simeq \mathrm{tot}(A^{\bullet,\bullet}) \to C_1^\bullet \otimes \cdots \otimes C_k^\bullet$$

is the given by the composition of the projection and the augmentation map, i.e.

$$P^0 = \left(\mathrm{tot}(A^{\bullet,\bullet})^0 = \bigoplus_{p+q=0} A^{p,q} \to A^{0,0} \xrightarrow{\otimes_{i=1}^{k} p_i^0} C_1^0 \otimes \cdots \otimes C_k^0 \right).$$

Thus, the map in degree zero on cohomology $\mathcal{H}^0(P)$ is given by the surjection $E^0 \to E_\infty^{0,0}$ followed by the morphism induced by $\otimes_{i=1}^{k} p_i^0$. This shows that $E_\infty^{0,0}$ equals $\mathcal{H}^0(C_1^\bullet \otimes \cdots \otimes C_k^\bullet)$ as a subsheaf of $C_1^0 \otimes \cdots \otimes C_k^0$. The assertion follows since

$$\mathcal{H}^0(C_1^\bullet \otimes \cdots \otimes C_k^\bullet) = \ker(d_1^0 \otimes \cdots \otimes d_k^0) = \mathrm{im}(\gamma_1 \otimes \cdots \otimes \gamma_k).$$

$\square$

Now theorem 2.6.1 is a corollary of theorem 2.7.1 and proposition 2.7.2.

Chapter 3

Cohomological invariants of twisted products of tautological sheaves

In this whole chapter let $E_1, \ldots, E_k$ be locally free sheaves on a quasi-projective surface X and $n \in \mathbb{N}$.

3.1 Description of $p_* q^* (E_1^{[n]} \otimes \cdots \otimes E_k^{[n]})$

We will use proposition 2.6.1 in order to study the image of $E_1^{[n]} \otimes \cdots \otimes E_k^{[n]}$ under the Bridgeland–King–Reid equivalence. We have to keep track of the $\mathfrak{S}_n$-linearization, since we later want compute the $\mathfrak{S}_n$-invariant cohomology. In the case that $E_1 = \cdots = E_k = E$ we also have to keep track of the $\mathfrak{S}_k$-action in order to get later also results for the symmetric and wedge products of tautological bundles (see remark 2.6.2).

3.1.1 Construction of the T_ℓ and φ_ℓ

Definition 3.1.1. Let $k, n \in \mathbb{N}$. For $1 \le \ell \le k$ we define I_ℓ as the set of tuples of the form $(M; i, j; a)$ consisting of a subset $M \subset [k]$ with $|M| = \ell$, two numbers $i, j \in [n]$ with $i < j$, and a multi-index $a \colon [k] \setminus M \to [n]$. Given such a tuple we set

$$\hat{M} := \hat{M}(i, j; a) := M \cup a^{-1}(\{i, j\})$$

and $a_| := a_{|[k] \setminus \hat{M}}$. The data of $i, j \in [n]$ with $i < j$ is the same as the subset $\{i, j\} \subset [n]$. Thus we will also write $(M; \{i, j\}; a)$ instead of $(M; i, j; a)$.

For $E_1, \ldots, E_k$ locally free sheaves on X and $\ell = 1, \ldots, k$ we define the coherent sheaf

$$T_\ell(E_1, \ldots, E_k) := \bigoplus_{(M; i, j; a) \in I_\ell} S^{\ell - 1} N_{\Delta_{i,j}}^{\vee} \otimes \left(\bigotimes_{\alpha \in M} E_\alpha \right)_{i,j} \otimes \left(\bigotimes_{\beta \in [k] \setminus M} \mathrm{pr}_{a(\beta)}^* E_\beta \right)$$

on X^n. We will often leave out the E_i in the argument of T_ℓ and denote the direct summands by $T_\ell(M; i, j; a)$. If we want to emphasise the values of k and n we will put them in the left under respectively upper index of the objects and morphisms, e.g. we will write ${}^n_k T_\ell$. We can rewrite the summands as

$$T_\ell(M; i, j; a) = \left(S^{\ell-1}\Omega_X \otimes \left(\bigotimes_{\alpha \in \hat{M}} E_\alpha \right) \right)_{i,j} \otimes \left(\bigotimes_{\beta \in [k] \setminus \hat{M}} \mathrm{pr}^*_{a(\beta)} E_\beta \right)$$

or as (see [Har77, Chapter II 8])

$$T_\ell(M; i, j; a) = S^{\ell-1}(\mathcal{I}_{i,j}/\mathcal{I}_{i,j}^2) \otimes \left(\bigotimes_{\alpha \in \hat{M}} E_\alpha \right)_{i,j} \otimes \left(\bigotimes_{\beta \in [k] \setminus \hat{M}} \mathrm{pr}^*_{a(\beta)} E_\beta \right)$$

$$\cong (\mathcal{I}_{i,j}^{\ell-1}/\mathcal{I}_{i,j}^{\ell}) \otimes \left(\bigotimes_{\alpha \in \hat{M}} E_\alpha \right)_{i,j} \otimes \left(\bigotimes_{\beta \in [k] \setminus \hat{M}} \mathrm{pr}^*_{a(\beta)} E_\beta \right).$$

As in subsection 2.4 for the terms F_I, we get for $\sigma \in \mathfrak{S}_n$ by flat base change canonical isomorphisms

$$\sigma_* : T_\ell(M; \sigma^{-1}(\{i, j\}); \sigma^{-1} \circ a) \to \sigma^* T_\ell(M; i, j; a).$$

Thus, there is a $\mathfrak{S}_n$-linearization λ of T_ℓ given on local sections $s \in T_\ell$ by

$$\lambda_\sigma(s)(M; i, j; a) = \varepsilon^\ell_{\sigma, \sigma^{-1}(\{i,j\})} \sigma_* s(M; \sigma^{-1}(\{i, j\}); \sigma^{-1} \circ a).$$

Remark 3.1.2. For $\sigma = (i\ j)$ the map $\sigma_* : N_{\Delta_{ij}} \to N_{\Delta_{ij}}$ is given by multiplication with -1 (see lemma 1.5.20). Thus, $\sigma_* : T_\ell(M; i, j; a) \to T_\ell(M; i, j; \sigma^{-1} \circ a)$ is given by multiplication with $(-1)^{\ell-1}$. Together with the sign $\varepsilon^\ell_{\sigma, \sigma^{-1}(\{i,j\})}$ this makes σ act by -1 on $T_\ell(M; i, j; a)$ for every tuple $(M; i, j; a)$ such that $a^{-1}(\{i, j\}) = \emptyset$, i.e. if $\hat{M} = M$ (see also remark 2.4.2).

If $E_1 = \cdots = E_k$ we define a $\mathfrak{S}_k$-action on T_ℓ by

$$(\mu \cdot s)(M; i, j; a) := \mu \cdot s(\mu^{-1}(M); i, j; a \circ \mu)$$

for $\mu \in \mathfrak{S}_k$. The action of μ on the right-hand side is given by permuting the factors E_t of the tensor product. Since the two linearizations commute, they give a $\mathfrak{S}_n \times \mathfrak{S}_k$-linearization of T_ℓ. We will now successively define $\mathfrak{S}_n$- respectively $\mathfrak{S}_n \times \mathfrak{S}_k$-equivariant morphisms $\varphi_\ell : K_{\ell-1} \to T_\ell$, where

$$K_0(E_1, \ldots, E_k) := K_0 := \bigotimes_{t=1}^{k} C^0_{E_t} = \bigoplus_{a: [k] \to [n]} K_0(a) \quad , \quad K_0(a) = \bigotimes_{t=1}^{k} \mathrm{pr}^*_{a(t)} E_t$$

and $K_\ell := \ker(\varphi_\ell)$ for $\ell = 1, \ldots, k$. We set $I_0 := \mathrm{Map}([k], [n])$. We consider K_0 with the $\mathfrak{S}_n$-linearization λ given by $\lambda_\sigma(s)(a) := \sigma_* s(\sigma^{-1} \circ a)$ and, if all the E_t are equal, with the $\mathfrak{S}_k$-action $(\mu \cdot s)(a) := \mu \cdot s(a \circ \mu)$ (see also remark 2.6.2). For $(M; i, j; a) \in I_\ell$ we set

$$I_0 \supset I(M; i, j; a) := \big\{ c \colon [k] \to [n] \mid c(M) \subset \{i, j\},\, c_{|[k] \setminus M} = a \big\} = \big\{ a \uplus b \mid b \colon M \to \{i, j\} \big\}$$

and define $K_{\ell-1}(M; i, j; a)$ as the image of $K_{\ell-1}$ under the projection

$$K_{\ell-1} \hookrightarrow K_0 = \bigoplus_{c \in I_0} K_0(c) \to \bigoplus_{c \in I(M; i, j; a)} K_0(c)$$

We define the component $\varphi_\ell(M; i, j; a) \colon K_{\ell-1} \to T_\ell(M; i, j; a)$ of φ_ℓ as the composition of the projection $K_{\ell-1} \to K_{\ell-1}(M; i, j; a)$ with a morphism $K_{\ell-1}(M; i, j; a) \to T_\ell(M; i, j; a)$. We denote the latter morphism again by $\varphi_\ell(M; i, j; a)$ and will define it in the following. For $s \in K_{\ell-1} \subset K_0$ the above means that $s(c)$ for $c \notin I(M; i, j; a)$ does not contribute to $\varphi_\ell(s)(M; i, j; a)$. We assume first that for all $t \in \hat{M} = M \cup a^{-1}(\{i, j\})$ the bundles E_t equal the trivial line bundle, i.e. $E_t = \mathcal{O}_X$. Then for all $b \colon M \to \{i, j\}$ we have

$$K_0(a \uplus b) = H \quad , \quad T_\ell(M; i, j; a) = (\mathcal{I}_{i,j}^{\ell-1} / \mathcal{I}_{i,j}^\ell) \otimes H \quad , \quad H := \bigotimes_{t \in [k] \setminus \hat{M}} \mathrm{pr}_{a(t)}^\star E_t .$$

Thus, for a local section $s \in K_{\ell-1}$ the components $s(a \uplus b) \in K_{\ell-1}(M; i, j; a)$ are all sections of the same locally free sheaf H and we can define

$$\varphi_\ell(M; i, j; a)(s) := \sum_{b \colon M \to \{i, j\}} \varepsilon_b s(a \uplus b) \quad \mathrm{mod}\ \mathcal{I}_{ij}^\ell$$

where $\varepsilon_b = (-1)^{\#\{t | b(t) = j\}}$. Inductively the map $\varphi_\ell(M; i, j; a)$ is well defined, which means that $\varphi_\ell(M; i, j; a)(s) \in \mathcal{I}_{i,j}^{\ell-1} \cdot H$, since if we take any $m \in M$ we have

$$\varphi_\ell(M; i, j; a)(s) = \sum_{b \colon M \setminus \{m\} \to \{i, j\}} \varepsilon_b s(a \uplus b, m \mapsto i) - \sum_{b \colon M \setminus \{m\} \to \{i, j\}} \varepsilon_b s(a \uplus b, m \mapsto j) .$$

Both sums occuring are elements of $\mathcal{I}_{i,j}^{\ell-1}$ since because of $s \in \ker \varphi_{\ell-1}$ we have

$$\varphi_{\ell-1}(s)(M \setminus \{m\}; i, j; a, m \mapsto i) = \varphi_{\ell-1}(s)(M \setminus \{m\}; i, j; a, m \mapsto j) = 0 \quad \mathrm{mod}\ \mathcal{I}_{i,j}^{\ell-1} .$$

Let now E_t for $t \in \hat{M}$ be the trivial vector bundle of rank r_t. Let $r := \prod_{t \in \hat{M}} r_t$. Then for $b \colon M \to \{i, j\}$ we have $K_0(a \uplus b) = (\oplus_\alpha H)$ and $T_\ell = (\mathcal{I}_{ij}^{\ell-1} / \mathcal{I}_{ij}^\ell) \otimes (\oplus_\alpha H)$. Here the index α goes through all multi-indices $(\alpha_t | t \in \hat{M})$ with $1 \leq \alpha_t \leq r_t$. Now we can define $\varphi_\ell(M; i, j; a)$ the same way as before. The components $\varphi_\ell(M; i, j; a)(\alpha, \alpha')$ are zero if $\alpha \neq \alpha'$ and coincide with the $\varphi_\ell(M; i, j; a)$ from the trivial line bundle case if $\alpha = \alpha'$.

Remark 3.1.3. Every collection of automorphisms of the $E_t = \mathcal{O}_X^{\oplus r_t}$ for $t \in \hat{M}$ induces canonically an automorphism of $\oplus_\alpha H$. The morphism $\varphi_\ell(M; i, j; a)$ commutes with the automorphisms induced by this automorphism on its domain and codomain.

This observation allows us to define $\varphi_\ell(M; i, j; a)$ in the case of general locally free sheaves as follows. We choose an open covering $\{U_m\}_m$ of X such that on every open set U_m all the E_t are simultaneously trivial, say with trivialisations $\mu_{m,t} \colon E_{t|U_m} \xrightarrow{\cong} \mathcal{O}_{U_m}^{r_t}$. Then the tivialisations μ_{m_t} for $t \in \hat{M}$ induce over $\mathrm{pr}_{ij}^{-1}(U_m \times U_m)$ isomorphisms $K_0(a \uplus b) \cong (\oplus_\alpha H)$ and $T_\ell \cong (\mathcal{I}_{ij}^{\ell-1}/\mathcal{I}_{ij}^\ell) \otimes (\oplus_\alpha H)$. We define the restriction of $\varphi_\ell(M; i, j; a)$ to $\mathrm{pr}_{ij}^{-1}(U_m^2)$ under these isomorphisms as the morphism $\varphi_\ell(M; i, j; a)$ from the case of trivial vector bundles. It is independent of the choosen trivialisations by the above remark. Thus, the $\varphi_\ell(M; i, j; a)$ defined over the $\mathrm{pr}_{ij}^{-1}(U_m^2)$ for varying m glue together. Since the $\mathrm{pr}_{ij}^{-1}(U_m^2)$ cover the partial diagonal Δ_{ij}, which is the support of $T_\ell(M; i, j; a)$, this defines $\varphi_\ell(M; i, j; a)$ globally. Using lemma 1.2.1 one can check that the morphisms φ_ℓ are indeed equivariant.

3.1.2 The open subset X_{**}^n

As done in [Dan00] and [Sca09a], we consider the following open subvarieties of X^n, $S^n X$, and $X^{[n]}$. Let $W \subset S^n X$ be the closed subvariety of unordered tuples $\sum_{i=1}^n x_i$ with the property that $|\{x_1, \ldots, x_n\}| \leq n - 2$, i.e

$$W = \pi \left(\bigcup_{|J| = |K| = 2, J \neq K} \Delta_J \cap \Delta_K \right).$$

We set $S^n X_{**} := S^n X \setminus W$, $X_{**}^n := \pi^{-1}(S^n X_{**})$, $X_{**}^{[n]} := \mu^{-1}(S^n X_{**})$ and

$$I^n X_{**} = q^{-1}(X_{**}^{[n]}) = p^{-1}(X_{**}^n) = (X_{**}^{[n]} \times_{S^n X_{**}} X_{**}^n)_{\mathrm{red}}.$$

In summary, there is the following open immersion of commutative diagrams

$$
\begin{array}{ccc}
I^n X_{**} & \xrightarrow{p_{**}} & X_{**}^n \\
q_{**} \downarrow & & \downarrow \pi_{**} \\
X_{**}^{[n]} & \xrightarrow{\mu_{**}} & S^n X_{**}
\end{array}
\qquad \xhookrightarrow{\iota} \qquad
\begin{array}{ccc}
I^n X & \xrightarrow{p} & X^n \\
q \downarrow & & \downarrow \pi \\
X^{[n]} & \xrightarrow{\mu} & S^n X
\end{array}
$$

where we denote every open immersion $(\_)_{**} \hookrightarrow (\_)$ by ι. For a sheaf or complex of sheaves F on X^n, $S^n X$, $X^{[n]}$ or $I^n X$ we write F_{**} for its restriction to the appropriate open subset.

The codimensions of the complements are at least two. More precisely, we have

$$\mathrm{codim}(X^n \setminus X^n_{**}, X^n) = \mathrm{codim}(S^n X \setminus S^n X_{**}, S^n X) = 4\,,$$
$$\mathrm{codim}(X^{[n]} \setminus X^{[n]}_{**}, X^{[n]}) = \mathrm{codim}(I^n X \setminus I^n X_{**}, I^n X) = 2\,.$$

Lemma 3.1.4. *Let $E_1, \ldots, E_k$ be locally free sheaves on X. Then on X^n there is a natural isomorphism*

$$\iota_* \iota^* p_* q^*(E_1^{[n]} \otimes \cdots \otimes E_k^{[n]}) \cong p_* q^*(E_1^{[n]} \otimes \cdots \otimes E_k^{[n]})\,.$$

Proof. We apply lemma 1.5.26 with $f = p$ and $U = X^n_{**}$. $\qquad\square$

3.1.3 Description of $p_* q^*(E_1^{[n]} \otimes \cdots \otimes E_k^{[n]})_{**}$

Proposition 3.1.5. *Let $E_1, \ldots, E_k$ be locally free sheaves on X. Then on the open subset $X^n_{**} \subset X^n$ there is the equality*

$$K_{k**} = p_* q^*(E_1^{[n]} \otimes \cdots \otimes E_k^{[n]})_{**}$$

of subsheaves of K_{0**}.

Proof. We will often drop the indices $(\_)_{**}$ in this proof. Using proposition 2.6.1 it suffices to show that $K_k = \mathrm{im}(\gamma_{E_1} \otimes \cdots \otimes \gamma_{E_k})$. For fixed $1 \le i \le j \le n$ and $\ell \in [k]$ we denote by $\varphi_\ell(i,j)$ the direct sum of all $\varphi_\ell(M; i, j; a)$ with $(M; i, j; a) \in I_\ell$. On the open subset $X^n_{**} \subset X^n$ the pairwise diagonals $\Delta_{i,j}$ do not intersect. We denote the big diagonal by $\mathbb{D} = \cup_{1 \le i < j \le n} \Delta_{i,j}$. Then X^n_{**} is covered by the open subsets $V_{i,j} := (X^n_{**} \setminus \mathbb{D}) \cup \Delta_{i,j}$. Thus we can test the equality on the $V_{i,j}$ where

$$K_\ell = K_\ell(i,j) = \cap_{\ell=1}^k \ker \varphi_\ell(i,j)$$

holds. We will assume without loss of generality the case that $i = 1$ and $j = 2$. We consider as in the construction of the φ_ℓ an open covering $\{U_m\}_m$ of X on which all the E_t are simultaneously trivial. Since both K_k and $\mathrm{im}(\gamma_{E_1} \otimes \cdots \otimes \gamma_{E_k})$ equal K_0 on $V_{12} \setminus \Delta_{12} = X^n_{**} \setminus \mathbb{D}$, it is sufficient to show the equality on every member of the covering of Δ_{12} given by

$$U_m \times U_m \times U_{m_3} \times \cdots \times U_{m_n}\,.$$

Since on these open sets the maps $\varphi_\ell(1,2)$ are defined as the maps $\varphi_\ell(1,2)(\mathcal{O}_X^{\oplus r_1}, \ldots, \mathcal{O}_X^{\oplus r_k})$ under the trivializations, we may assume that all the E_t are trivial vector bundles of rank r_t, i.e. $E_t = \mathcal{O}_X^{\oplus r_t}$. Since in this case the φ_ℓ as well as $\gamma_{E_1} \otimes \cdots \otimes \gamma_{E_k}$ are defined component-wise, we may assume that $E_1 = \cdots = E_t = \mathcal{O}_X$. By theorem 2.5.3 a section $x \in p_* q^* \mathcal{O}_X^{[n]} \subset C^0_{\mathcal{O}_X}$ over V_{12} is of the form $x = (x(1), x(2), \ldots, x(n))$ with $x(\alpha) \in \mathrm{pr}_\alpha^* \mathcal{O}_X \cong \mathcal{O}_{X^n}$ for $\alpha \in [n]$ and $x(1)_{|\Delta_{12}} = x(2)_{|\Delta_{12}}$. For a section $s \in K_0$ and a multi-index $a \colon [k] \to [n]$ we denote the component of s in

$$K_0(a) = \mathrm{pr}_{a(1)}\,\mathcal{O}_X \otimes \cdots \otimes \mathrm{pr}_{a(k)}\,\mathcal{O}_X \cong \mathcal{O}_{X^n}$$

by $s(a)$. The image of a pure tensor $x_1 \otimes \cdots \otimes x_k \in (p_*q^*\mathcal{O}_X)^{\otimes k}$ under the k-th power of $\gamma = \gamma_{\mathcal{O}_X}$ is given by

$$\gamma^{\otimes k}(x_1 \otimes \cdots \otimes x_k)(a) = x_1(a(1)) \cdots x_k(a(k)) \in \mathcal{O}_{X^n}\,.$$

For a tuple (M,a) with $\emptyset \neq M \subset [k]$, $a\colon [k] \setminus M \to [n]$, and $s \in K_0$ we set

$$s(M,a) = \sum_{b\colon M \to [2]} \varepsilon_b s(a \uplus b) \in \mathcal{O}_{X^n}\,.$$

Then for a section $s \in K_0$ being a section of $K_k = K_k(1,2)$ is equivalent to the condition that $s(M,a) \in \mathcal{I}^{|M|}$ for each tupel (M,a) as above, where $\mathcal{I} := \mathcal{I}_{12}$. We first show the inclusion $\mathrm{im}(\gamma^{\otimes k}) \subset K_k$. For this let $x = x_1 \otimes \cdots \otimes x_k \in (p_*q^*\mathcal{O}_X)^{\otimes k}$ and $s = \gamma^{\otimes k}(x)$. We show by induction over $|M|$ that $s(M,a) \in \mathcal{I}^{|M|}$ for each pair (M,a). For $M = \{t\}$ we have

$$s(\{t\},a) = s(t \mapsto 1, a) - s(t \mapsto 2, a) = (x_t(1) - x_t(2)) \cdot \prod_{i \in [k] \setminus \{t\}} x_i(a(i))$$

which is indeed a section of $\mathcal{I}$ since $x_t(1)_{|\Delta_{12}} = x_t(2)_{|\Delta_{12}}$. For an arbitrary $M \subset [k]$ we choose an $m \in M$ and set

$$\tilde{x} = x_1 \otimes \cdots \otimes x_{m-1} \otimes \tilde{x}_m \otimes x_{m+1} \otimes \cdots \otimes x_k$$

with $\tilde{x}_m(j) = 1$ for every $j \in [n]$. We also set $\tilde{s} = \gamma^{\otimes k}(\tilde{x})$. With this notation

$$s(M,a) = (x_m(1) - x_m(2)) \cdot \tilde{s}(M \setminus \{k\}, a, m \mapsto 1)\,.$$

By induction we have $\tilde{s}(M \setminus \{k\}, a, m \mapsto 1) \in \mathcal{I}^{|M|-1}$ and thus $s(M,a) \in \mathcal{I}^{|M|}$. For the inclusion $K_k \subset \mathrm{im}\,\gamma^{\otimes k}$ we need the following lemma, where we are still working over V_{12}. For $a\colon [k] \to [n]$ we set $\hat{M}(a) := a^{-1}(\{1,2\})$ and $a_| := a_{|[k] \setminus \hat{M}(a)}$.

Lemma 3.1.6. *Let $s \in K_k$ be a local section and $a\colon [k] \to [n]$ such that $s(b) = 0$ for all $b\colon [k] \to [n]$ with $(\hat{M}(b), b_|) = (\hat{M}(a), a_|)$ and $|b^{-1}(\{2\})| < |a^{-1}(\{2\})|$. Then there exists a local section $x \in (p_*q^*\mathcal{O}_X^{[n]})^{\otimes k}$ such that $\gamma^{\otimes k}(x)(a) = s(a)$ and $\gamma^{\otimes k}(x)(c) = 0$ for all multi-indices $c\colon [k] \to [n]$ with the property that $(\hat{M}(c), c_|) \neq (\hat{M}(a), a_|)$ or with the property that there exists an $i \in \hat{M}(a) = \hat{M}(c)$ with $c(i) = 1$ and $a(i) = 2$.*

Proof. We assume for simplicity that $\hat{M}(a) = [u]$ and $a^{-1}(\{2\}) = [v]$ with $1 \leq v \leq u \leq k$, i.e. a is of the form

$$a = (2, \ldots, 2, 1, \ldots, 1, a(u+1), \ldots, a(k))\,.$$

By the assumptions

$$\mathcal{I}^v \ni s([v], a_{|[v+1,k]}) = s([v], \underline{1} \uplus a_|) = \sum_{b:\, [v] \to [2]} \varepsilon_b s(b \uplus \underline{1} \uplus a_|) = (-1)^v s(a).$$

The last equality is because of $\hat{M}(b \uplus \underline{1} \uplus a_|) = \hat{M}(a)$ and $(b \uplus \underline{1} \uplus a_|)_| = a_|$ for all $b: [v] \to [2]$. Now we can write $s(a) = \sum_{\alpha \in A} y_{\alpha,1} \cdots y_{\alpha,v}$ as a finite sum with all $y_{\alpha,\beta} \in \mathcal{I}$. We denote by e_j the section of $C^0_{\mathcal{O}_X}$ with $e_j(h) = \delta_{jh}$. Then the section

$$x = \sum_{\alpha \in A} y_{\alpha,1} e_2 \otimes \cdots \otimes y_{\alpha,v} e_2 \otimes (e_1 + e_2) \otimes \cdots \otimes (e_1 + e_2) \otimes e_{a(u+1)} \otimes \cdots \otimes e_{a(k)}$$

is indeed in $(p_* q^* \mathcal{O}_X^{[n]})^{\otimes k}$ and has the desired properties. □

Let $\prec$ be any total order on the set of tuples (M, a) with $M \subset [k]$ and $a: [k] \setminus M \to [3, n]$ and let $\lhd_{M,v}$ be any total order on the set of subsets of M of cardinality v. We define a total order $<$ on the set $I_0 = \mathrm{Map}([k], [n])$ by setting $b < a$ if $(\hat{M}(b), b_|) \prec (\hat{M}(a), a_|)$ or if $(\hat{M}(b), b_|) = (\hat{M}(a), a_|)$ and $|b^{-1}(2)| < |a^{-1}(2)|$ or if $(\hat{M}(b), b_|) = (\hat{M}(a), a_|) =: M$ and $|b^{-1}(2)| = |a^{-1}(2)| =: v$ and $b^{-1}(2) \lhd_{M,v} a^{-1}(2)$. For $s \in K_k$ let a be the minimal multi-index with $s(a) \neq 0$, i.e. $s(b) = 0$ for all $b < a$. Then lemma 3.1.6 yields a $x \in (p_* q^* \mathcal{O}_X^{[n]})^{\otimes k}$ such that $\hat{s} = s - \gamma^{\otimes k}(x)$ fulfills $\hat{s}(b) = 0$ for all $b \leq a$. Thus, by induction over the set I_0 with the order $<$, indeed, $s \in \mathrm{im}(\gamma^{\otimes k})$ which completes the proof of proposition 3.1.5. □

3.1.4 Description of $p_* q^*(E_1^{[n]} \otimes \cdots \otimes E_k^{[n]})$

The result of the last subsection carries over directly to the whole X^n.

Theorem 3.1.7. Let $E_1, \ldots, E_k$ be *locally free sheaves on X. Then on X^n there is the equality*

$$K_k = p_* q^*(E_1^{[n]} \otimes \cdots \otimes E_k^{[n]})$$

of subsheaves of K_0.

Proof. Since K_0 is locally free and $\mathrm{codim}(X^n \setminus X^n_{**}, X^n) = 4$ we have $\iota_* K_{0**} = K_0$. Furthermore the direct summands of T_ℓ are push forwards of locally free sheaves on the partial diagonals $\Delta_{i,j}$. Since

$$\mathrm{codim}(\Delta_{ij} \setminus (\Delta_{ij} \cap X^n_{**}), \Delta_{i,j}) = 2$$

we get by lemma 1.5.26 that $\iota_* T_{\ell**} = T_\ell$ for all $\ell \in [k]$. Using lemma 1.5.27 we get by induction that $\iota_* K_{\ell**} = K_\ell$ for $\ell \in [k]$. In particular

$$K_k = \iota_* K_{k**} \overset{3.1.5}{=} \iota_* \left(p_* q^*(E_1^{[n]} \otimes \cdots \otimes E_k^{[n]})_{**} \right) \overset{3.1.4}{=} p_* q^*(E_1^{[n]} \otimes \cdots \otimes E_k^{[n]}).$$

$\square$

Corollary 3.1.8. *There are natural isomorphisms* $\mu_*(E_1^{[n]} \otimes \cdots \otimes E_k^{[n]}) \cong K_k^{\mathfrak{S}_n}$ *and*

$$\mathrm{H}^*(X^{[n]}, E_1^{[n]} \otimes \cdots \otimes E_k^{[n]}) \cong \mathrm{H}^*(X^n, K_k)^{\mathfrak{S}_n}.$$

Proof. This follows by the previous theorem together with proposition 2.2.4 and corollary 2.5.2. $\square$

3.2 Invariants of K_0 and the T_ℓ

In this section we will use Danila's lemma (see section 1.5.2) in order to compute the invariants of the sheaves K_0 and T_ℓ.

3.2.1 Orbits and their isotropy groups on the sets of indices

For $\ell = 1, \ldots, k$ we have the decomposition $T_\ell = \oplus_{I_\ell} T_\ell(M; i, j; a)$ with

$$I_\ell = \{(M; i, j; a) \mid M \subset [k],\ \#M = \ell,\ 1 \le i < j \le n,\ a: [k] \setminus M \to [n]\}\,.$$

The $\mathfrak{S}_n$- as well as the $\mathfrak{S}_k$-linearization of T_ℓ induce actions on I_ℓ given for $\sigma \in \mathfrak{S}_n$ and $\mu \in \mathfrak{S}_k$ by

$$\sigma \cdot (M; \{i,j\}; a) = (M; \sigma(\{i,j\}); \sigma \circ a)\quad,\quad \mu \cdot (M; \{i,j\}; a) = (\mu(M); \{i,j\}; a \circ \mu^{-1})\,.$$

Furthermore there is the decomposition

$$K_0 = \bigoplus_{a \in I_0} K_0(a)\quad,\quad I_0 = \mathrm{Map}([k], [n])\quad,\quad K_0(a) = \bigotimes_{t=1}^{k} \mathrm{pr}_{a(t)}^* E_t\,.$$

The $\mathfrak{S}_n \times \mathfrak{S}_k$-action on I_0 is given by $\sigma \cdot a = \sigma \circ a$ and $\mu \cdot a = a \circ \mu^{-1}$. We define a total order $\prec$ on the set of subsets of $[k]$ by setting for $A, B \subset [k]$ two subsets

$$A \prec B : \iff \begin{cases} |A| > |B| \text{ or} \\ |A| = |B| \text{ and } \min(A \setminus B) < \min(B \setminus A). \end{cases}$$

Remark 3.2.1. Let $1 \le \ell \le k$.

(i) Every $\mathfrak{S}_n$-orbit of I_0 has a unique representative a such that

$$a^{-1}(1) \prec a^{-1}(2) \prec \cdots \prec a^{-1}(n)\,.$$

We denote the set of these representatives by $J_0(1)$. For $a \in I_0$ the isotropy group is given by $\mathrm{Stab}_{\mathfrak{S}_n}(a) = \mathfrak{S}_{[n]\setminus\mathrm{im}(a)}$. For $a \in J_0(1)$ we have $[n] \setminus \mathrm{im}(a) = [\max a + 1, n]$.

(ii) Every $\mathfrak{S}_n$-orbit of I_ℓ has a unique representative of the form $(M; 1, 2; a)$ such that $a^{-1}(1) \prec a^{-1}(2)$ and

$$a^{-1}(3) \prec a^{-1}(4) \prec \cdots \prec a^{-1}(n).$$

We denote the set of these representatives by $J_\ell(1)$. Furthermore, we set

$$\hat{I}_\ell := \left\{ (M; i, j; a) \in I_\ell \mid a^{-1}(\{i, j\}) \neq \emptyset \right\} \quad , \quad \hat{J}_\ell(1) := J_\ell(1) \cap \hat{I}_\ell.$$

We will often use the identification $(M; a) \cong (M; 1, 2; a) \in J_\ell(1)$ in the notations. The isotropy group of a tuple $(M; i, j; a) \in I_\ell$ with $Q := \{i, j\} \cup \mathrm{im}(a)$ and $\bar{Q} = [n] \setminus Q$ is given by

$$\mathrm{Stab}_{\mathfrak{S}_n}(M; i, j; a) = \begin{cases} \mathfrak{S}_{\bar{Q}} & \text{if } (M; i, j; a) \in \hat{I}_\ell, \\ \mathfrak{S}_{\{i,j\}} \times \mathfrak{S}_{\bar{Q}} & \text{if } (M; i, j; a) \notin \hat{I}_\ell. \end{cases}$$

If $(M; i, j; a) = (M; 1, 2; a) \in J_\ell(1)$ we have $Q = [\max(a, 2)]$ and $\bar{Q} = [\max(a, 2) + 1, n]$.

(iii) Every $\mathfrak{S}_k$-orbit of I_0 has a unique representative that is non-decreasing. We denote the set of these representatives by $J_0(2)$. For $a \in I_0$ the isotropy group is given by

$$\mathrm{Stab}_{\mathfrak{S}_k}(a) = \prod_{v=1}^{n} \mathfrak{S}_{a^{-1}(v)}.$$

(iv) Every $\mathfrak{S}_k$-orbit of I_ℓ has a unique representative of the form $([\ell]; i, j; a)$ such that a is non-decreasing when considering $[n]$ with the order

$$i < j < 1 < \cdots < \hat{i} < \cdots < \hat{j} < \cdots < n.$$

We denote the set of these representatives by $J_\ell(2)$. The isotropy group of a tuple $(M; i, j; a) \in I_\ell$ is given by

$$\mathrm{Stab}_{\mathfrak{S}_k}(M; i, j; a) = \mathfrak{S}_M \times \prod_{t=1}^{n} \mathfrak{S}_{a^{-1}(t)}.$$

(v) Every $\mathfrak{S}_n \times \mathfrak{S}_k$-orbit of I_0 has a unique representative a which is non-decreasing and such that $a^{-1}(1) \prec \cdots \prec a^{-1}(n)$. The isotropy group of $a \in I_0$ is given by

$$\mathrm{Stab}_{\mathfrak{S}_n \times \mathfrak{S}_k} = \left\{ (\sigma, \mu) \mid \sigma a \mu^{-1} = a \right\}.$$

(vi) Every $\mathfrak{S}_n \times \mathfrak{S}_k$-orbit of I_ℓ has a unique representative of the form $([\ell]; 1, 2; a)$ such that a is non-decreasing when considering $[n]$ with the order as in (iv) and such that $a^{-1}(1) \prec a^{-1}(2)$ and

$$a^{-1}(3) \prec a^{-1}(4) \prec \cdots \prec a^{-1}(n).$$

We denote the set of these representatives by $J_\ell(3)$. The isotropy group of a tuple $(M; i, j; a) \in I_\ell$ is given by

$$\mathrm{Stab}_{\mathfrak{S}_n \times \mathfrak{S}_k}(M; i, j; a) = \left\{ (\sigma, \mu) \mid \sigma(\{i, j\}) = \{i, j\},\ \mu(M) = M,\ \sigma a \mu^{-1} = a \right\}.$$

3.2.2 The sheaves of invariants and their cohomology

Lemma 3.2.2. *There is a natural isomorphism*

$$K_0^{\mathfrak{S}_n} \cong \bigoplus_{a \in J_0(1)} K_0(a)^{\mathfrak{S}_{[\max a + 1, n]}}.$$

Proof. This follows from Danila's lemma and remark 3.2.1 (i). $\qquad\square$

Lemma 3.2.3.
$$T_\ell^{\mathfrak{S}_n} \cong \bigoplus_{(M; a) \in \hat{J}_\ell(1)} T_\ell(M; a)^{\mathfrak{S}_{[\max(a,2)+1, n]}}$$

holds for every $\ell = 1, \ldots, k$.

Proof. By Danila's lemma and remark 3.2.1 (ii) we have

$$T_\ell^{\mathfrak{S}_n} = \bigoplus_{(M; a) \in J_\ell(1)} T_\ell(M; a)^{\mathrm{Stab}(M; a)}.$$

Let $(M; a) \in J_\ell(1) \setminus \hat{J}_\ell(1)$. Then $\tau = (1\ 2) \in \mathrm{Stab}(M; a)$ acts on $T_\ell(M; a)$ by $(-1)^{\ell + \ell - 1} = -1$ (see remark 3.1.2) which makes the invariants vanish. $\qquad\square$

Corollary 3.2.4. *The sheaf $T_k^{\mathfrak{S}_n}$ is zero and thus*

$$\mu_*(E_1^{[n]} \otimes \cdots \otimes E_k^{[n]}) \cong p_* q^*(E_1^{[n]} \otimes \cdots \otimes E_k^{[n]})^{\mathfrak{S}_n} \cong K_{k-1}^{\mathfrak{S}_n}.$$

Proof. The set $\hat{J}_k(1)$ is empty. The isomorphisms follow by corollary 3.1.8. $\qquad\square$

Remark 3.2.5. For a subset $Q \subset [n]$ with $|Q| = q$ and $\bar{Q} = [n] \setminus Q$ we denote by $X^Q \times S^{\bar{Q}} X$ the quotient of X^n by the $\mathfrak{S}_{\bar{Q}}$-action. It is isomorphic to $X^q \times S^{n-q} X$. We denote by $\pi_Q \colon X^Q \times S^{\bar{Q}} X \to S^n X$ the morphism induced by the quotient morphism $\pi \colon X^n \to S^n X$.

Under the identification $X^Q \times S^{\bar{Q}}X \cong X^q \times S^{n-q}$ it is given by

$$(x_1, \ldots, x_q, \Sigma) \mapsto x_1 + \cdots + x_q + \Sigma.$$

Let $a \in I_0$ respectively $(M; i, j; a) \in \hat{I}_\ell$ and $Q = \operatorname{im}(a)$ respectively $Q = \{i, j\} \cup \operatorname{im}(a)$, and $\bar{Q} = [n] \setminus Q$. The sheaves $K_0(a)^{\mathfrak{S}_Q}$ respectively $T_\ell(M; i, j; a)^{\mathfrak{S}_Q}$ in the two lemmas above are considered as sheaves on the $\mathfrak{S}_n$-quotient $S^n X$, i.e. they are abbreviations

$$K_0(a)^{\mathfrak{S}_Q} := (\pi_* K_0(a))^{\mathfrak{S}_Q} \quad , \quad T_\ell(M; i, j; a)^{\mathfrak{S}_Q} := (\pi_* T_\ell(M; i, j; a))^{\mathfrak{S}_Q}$$

But we can also take the $\mathfrak{S}_{\bar{Q}}$-invariants already on the $\mathfrak{S}_{\bar{Q}}$-quotient $X^Q \times S^{\bar{Q}}X$ and consider $K_0(a)^{\mathfrak{S}_Q}$ and $T_\ell(M; i, j; a)^{\mathfrak{S}_Q}$ as sheaves on this variety. With this notation we have

$$(\pi_* K_0(a))^{\mathfrak{S}_Q} = \pi_{Q*}(K_0(a)^{\mathfrak{S}_Q}) \quad , \quad (\pi_* T_\ell(M; i, j; a))^{\mathfrak{S}_Q} = \pi_{Q*}(T_\ell(M; i, j; a)^{\mathfrak{S}_Q}).$$

We denote for $m \in Q$ by $p_m \colon X^Q \times S^{\bar{Q}}X \to X$ the projection induced by the projection $\operatorname{pr}_m \colon X^n \to X$. For $I \subset Q$ we have the closed embedding $\Delta_I \times S^{\bar{Q}}X \subset X^Q \times S^{\bar{Q}}X$ which is the $\mathfrak{S}_{\bar{Q}}$-quotient of the closed embedding $\Delta_J \subset X^n$. Then the sheaves of invariants considered as sheaves on $X^Q \times S^{\bar{Q}}X$ are given by

$$K_0(a)^{\mathfrak{S}_Q} = \bigotimes_{\substack{m \in Q \\ t \in a^{-1}(m)}} p_m^* E_t \,, \; T_\ell(M; i, j; a)^{\mathfrak{S}_Q} = \Big(\bigotimes_{t \in M \cup a^{-1}(\{i,j\})} E_t \Big)_{ij} \otimes \bigotimes_{\substack{m \in Q \setminus \{i,j\} \\ t \in a^{-1}(m)}} p_m^* E_t \,.$$

In particular, $K_0(a)^{\mathfrak{S}_Q}$ is still locally free. The sheaf $T(M; i, j; a)^{\mathfrak{S}_Q}$ can also be considered as a sheaf on its support $\Delta_{ij} \times S^{\bar{Q}}X$ on which it is locally free, too.

For the following, remember that we interpret an empty tensor product of sheaves on the surface X as the structural sheaf $\mathcal{O}_X$.

Lemma 3.2.6. *(i) For every $a \in I_0$ the cohomology $\mathrm{H}^*(X^n, K_0(a))$ is naturally isomorphic to*

$$\mathrm{H}^* \Big(\bigotimes_{t \in a^{-1}(1)} E_t \Big) \otimes \cdots \otimes \mathrm{H}^* \Big(\bigotimes_{t \in a^{-1}(n)} E_t \Big).$$

(ii) For every $a \in J_0(1)$ the invariant cohomology $\mathrm{H}^*(X^n, K_0(a))^{\mathfrak{S}_{[\max a+1, n]}}$ is naturally isomorphic to

$$\mathrm{H}^* \Big(\bigotimes_{t \in a^{-1}(1)} E_t \Big) \otimes \cdots \otimes \mathrm{H}^* \Big(\bigotimes_{t \in a^{-1}(\max a)} E_t \Big) \otimes S^{n - \max a} \, \mathrm{H}^*(\mathcal{O}_X).$$

(iii) For every $(M; i, j; a) \in I_\ell$ the cohomology $\mathrm{H}^*(X^n, T_\ell(M; i, j; a))$ is naturally isomorphic

to

$$\mathrm{H}^* \left(S^{\ell-1}\Omega_X \otimes \bigotimes_{t\in \hat{M}(a)} E_t \right) \otimes \bigotimes_{m\in[n]\setminus\{i,j\}} \mathrm{H}^* \left(\bigotimes_{t\in a^{-1}(m)} E_t \right).$$

(iv) For $(M;a) \in \hat{J}_\ell(1)$ the invariant cohomology $\mathrm{H}^*\left(X^n, T_\ell(M;a)\right)^{\mathfrak{S}_{[\max(a,2)+1,n]}}$ is naturally isomorphic to

$$\mathrm{H}^* \left(S^{\ell-1}\Omega_X \otimes \bigotimes_{t\in \hat{M}(a)} E_t \right) \otimes \bigotimes_{m=3}^{\max a} \mathrm{H}^* \left(\bigotimes_{t\in a^{-1}(m)} E_t \right) \otimes S^{n-\max(a,2)}\, \mathrm{H}^*(\mathcal{O}_X).$$

Proof. The natural isomorphisms in (i) and (iii) are the Künneth isomorphisms. The assertions (ii) and (iv) follow from the fact that the natural $\mathfrak{S}_n$-linearization of $\mathcal{O}_{X^n}$ induces the action on

$$\mathrm{H}^*(X^n, \mathcal{O}_{X^n}) \cong \mathrm{H}^*(\mathcal{O}_X)^{\otimes n}$$

given by permuting the tensor factors together with the cohomoligcal sign $\varepsilon_{\sigma,p_1,\ldots,p_n}$ (see section 1.3). $\square$

Lemma 3.2.7. *(i) For every $a \in J_0(1)$ the Euler characteristic of the invariants of $K_0(a)$ is given by*

$$\chi\left(K_0(a)^{\mathfrak{S}_{[\max a+1,n]}}\right) = \prod_{m=1}^{\max a} \chi\left(\bigotimes_{t\in a^{-1}(m)} E_t \right) \cdot \binom{\chi(\mathcal{O}_X)+n-\max a - 1}{n-\max a}.$$

(ii) For every $(M;a) \in \hat{J}_\ell(1)$ the Euler characteristic $\chi(T_\ell(M;a)^{\mathfrak{S}_{[\max(a,2)+1,n]}})$ is given by

$$\chi\left(S^{\ell-1}\Omega_X \otimes \bigotimes_{t\in \hat{M}(a)} E_t \right) \cdot \prod_{m=3}^{\max a} \chi\left(\bigotimes_{t\in a^{-1}(m)} E_t \right) \cdot \binom{\chi(\mathcal{O}_X)+n-\max(a,2)-1}{n-\max(a,2)}.$$

Proof. This follows from the previous lemma and lemma 1.3.1. $\square$

3.3 The map φ_1 on cohomology and the cup product

We consider the morphism $\varphi_1\colon K_0 \to T_1$ defined in subsection 3.1.1 and for $a \in I_0$ and $(\{t\};i,j;b) \in I_1$ its components

$$\varphi_1(a \to (\{t\};i,j;b))\colon K_0(a) \to T_1(\{t\};i,j;b).$$

The morphism $\varphi_1(a \to (\{t\}; i, j; b))$ is non-zero only if $a_{|[k]\setminus\{t\}} = b$ and $a(t) \in \{i,j\}$. In this case it is given by $\varepsilon_{a(t),\{i,j\}}$ times the morphims given by restricting sections to the pairwise diagonal Δ_{ij}.

Remark 3.3.1. Let $s \in K_0$ be a local section such that $\varphi_1(s) = 0$. Then for all pairs $a, b \in I_0$ and distinct $i, j \in [n]$ such that $a^{-1}(\{i,j\}) = b^{-1}(\{i,j\})$ and $a_{|[k]\setminus a^{-1}(\{i,j\})} = b_{|[k]\setminus b^{-1}(\{i,j\})}$ we have $s(a)_{|\Delta_{ij}} = s(b)_{|\Delta_{ij}}$. This implies that for every global section $s \in K_1 = \ker(\varphi_1)$ its restriction $s_{|\Delta_{ij}}$ is $(i\ j)$-invariant. Indeed, the action is given by $((i\ j)\cdot s)(a) = (i\ j)_* s((i\ j) \circ a)$. Now, $(i\ j)_*$ is the identity on Δ_{ij} and the pair $a, (i\ j) \circ a$ has the properties of the pair a, b form above.

For two sheaves $F, G \in \mathrm{Coh}(X)$ (or more generally two objects in $\mathrm{D}^b(X)$) the composition

$$\mathrm{H}^*(X, F) \otimes \mathrm{H}^*(X, G) \cong \mathrm{H}^*(X^2, F \boxtimes G) \to \mathrm{H}^*(X, F \otimes G)$$

of the Künneth isomorphism and the map induced by the restriction to the diagonal equals the cup product. Thus the map $H^*(\varphi_1(a, (\{t\}; i, j; a_{|[k]\setminus\{t\}})))$ is given in terms of the natural isomorphisms of lemma 3.2.6 by sending a

$$v_1 \otimes \cdots \otimes v_n \in \mathrm{H}^*\left(\bigotimes_{t \in a^{-1}(1)} E_t\right) \otimes \cdots \otimes \mathrm{H}^*\left(\bigotimes_{t \in a^{-1}(n)} E_t\right)$$

to the class

$$(v_i \cup v_j) \otimes v_1 \otimes \cdots \otimes \hat{v}_i \otimes \cdots \otimes \hat{v}_j \otimes \cdots \otimes v_n \in \mathrm{H}^*\left(\bigotimes_{t \in a^{-1}(\{i,j\})} E_t\right) \otimes \bigotimes_{m \in [n]\setminus\{i,j\}} \mathrm{H}^*\left(\bigotimes_{t \in a^{-1}(m)} E_t\right).$$

Remember (section 2.6) that there are the augmentation morphisms $\gamma_i \colon p_* q^* E_i^{[n]} \to C_{E_i}^0$ and that $K_0 = \otimes_{i=1}^k C_{E_i}^0$. We consider the composition

$$\bigotimes_{i=1}^k \mathrm{H}^*(X^n, p_* q^* E_i^{[n]}) \xrightarrow{\cup} \mathrm{H}^*(X^n, \bigotimes_{i=1}^k p_* q^* E_i^{[n]}) \xrightarrow{\mathrm{H}^*(\otimes_i \gamma_i)} \mathrm{H}^*(X^n, K_0).$$

Taking (factor-wise) the $\mathfrak{S}_n$-invariants we get the map (see proposition 2.5.5)

$$\Psi \colon \bigotimes_{i=1}^k \mathrm{H}^*(X^{[n]}, E_i^{[n]}) \cong \bigotimes_{i=1}^k \left(\mathrm{H}^*(E_i) \otimes S^{n-1} \mathrm{H}^*(\mathcal{O}_X)\right) \to \mathrm{H}^*(X^n, K_0)^{\mathfrak{S}_n}.$$

This map coincides with the $\mathfrak{S}_n$-invariant cup product $\otimes_i \mathrm{H}^*(C_{E_i}^0)^{\mathfrak{S}_n} \to \mathrm{H}^*(\otimes_i C_{E_i}^0)^{\mathfrak{S}_n}$. The inclusion $p_* q^* (\otimes_i E_i^{[n]})^{\mathfrak{S}_n} \subset K_0^{\mathfrak{S}_n}$ induces a map $\mathrm{H}^*(X^{[n]}, \otimes_i E_i^{[n]}) \to \mathrm{H}^*(X^n, K_0)^{\mathfrak{S}_n}$. Since $\mathrm{im}(\otimes_i \gamma_i) = \otimes_i p_* q^* E_i^{[n]}$ (proposition 2.6.1), the image of Ψ is a subset of the image of this

64

map. In degree zero the map $H^0(X^{[n]}, \otimes_i E_i^{[n]}) \to H^0(X^n, K_0)^{\mathfrak{S}_n}$ is a inclusion. Thus, we have $\mathrm{im}(\Psi) \subset H^0(X^{[n]}, \otimes_i E_i^{[n]})$. Let X be projective. In this case $H^0(\mathcal{O}_X) = \langle 1 \rangle \cong \mathbb{C}$ where 1 is the function with constant value one. Thus, we have for $a \in I_0$ the formula (see lemma 3.2.6)

$$H^0(X^n, K_0(a)) \cong \bigotimes_{m \in \mathrm{im}(a)} H^0 \left(\bigotimes_{r \in a^{-1}(m)} E_r \right)$$

and the action of $\mathrm{Stab}_{\mathfrak{S}_n}(a) = \mathfrak{S}_{\overline{\mathrm{im}\,a}}$ on this vector spaces is the trivial one, which means $H^0(X^n, K_0(a))^{\mathfrak{S}_{\overline{\mathrm{im}\,a}}} = H^0(X^n, K_0(a))$. Now, for

$$x_i \in H^0(E_i^{[n]}) \cong H^0(E_i) \otimes S^{n-1} H^0(\mathcal{O}_X) \cong H^0(E_i)$$

and $a \in J_0(1)$ we have

$$\Psi(x_1 \otimes \cdots \otimes x_k)(a) = \bigotimes_{m \in \mathrm{im}(a)} (\cup_{r \in a^{-1}(m)} x_r).$$

3.4 Cohomology in the highest and lowest degree

3.4.1 Global sections for $n \geq k$ and X projective

In this subsection we assume that X is projective. We will generalise the formula given in [Dan07] for the global sections of tensor powers of a tautological sheaf associated to a line bundles to a formula for tensor products of arbitrary tautological bundles.

Lemma 3.4.1. *Let $a\colon [k] \to [\ell]$, $t \in [k]$, $i = a(t)$, and $j \in [n] \setminus \mathrm{im}(a)$. Then under the natural isomorphisms of lemma 3.2.6 the map $H^0(\varphi_1)(a \to (\{t\}; \{i,j\}; a_{|[k]\setminus\{t\}}))$ corresponds to $\varepsilon_{a(t),\{i,j\}}$ times the identity on $\otimes_{m \in \mathrm{im}(a)} H^0(\otimes_{r \in a^{-1}(m)} E_r)$.*

Proof. This follows by the formula for $H^*(\varphi_1)(a \to (\{t\}; \{i,j\}; a_{|[k]\setminus\{t\}}))$ of section 3.3 and the fact that $v \cup 1 = v$ for every $v \in H^0(\otimes_{r \in a^{-1}(i)} E_r)$. $\square$

Lemma 3.4.2. *Let $m = \min(n,k)$. Then every $s \in \ker H^0(\varphi_1)$ is determined by its components $s(a) \in H^0(K_0(a))$ for $a \in I_0$ with $|\mathrm{im}(a)| = m$.*

Proof. We use induction over $w := m - \mathrm{im}(b)$ with the hypothesis that $s(b)$ is determined by the values of the $s(a)$ with $|\mathrm{im}(a)| = m$. Clearly, the hypothesis is true for $w = 0$. So now let $b\colon [n] \to [k]$ with $|\mathrm{im}(b)| < \min(n,k)$. Such a map b is neither injective nor surjective. Thus, we can choose $j \in [n] \setminus \mathrm{im}(a)$ and a pair $t, t' \in [k]$ with $t \neq t'$ and $i := b(t) = b(t')$. We define $\tilde{b}\colon [k] \to [n]$ by $\tilde{b}_{|[k]\setminus\{t\}} = b_{|[k]\setminus\{t\}}$ and $\tilde{b}(t) = j$. Then $\mathrm{im}(\tilde{b}) = \mathrm{im}(b) \cup \{j\}$ which gives

$|\mathrm{im}(\tilde{b})| = |\mathrm{im}(b)| + 1$. We have

$$
\begin{aligned}
0 &= \mathrm{H}^0(\varphi_1)(s)(\{t\}; \{i,j\}; b_{|[k]\setminus\{t\}}) \\
&= \mathrm{H}^0(\varphi_1)(b \to (\{t\}; \{i,j\}; b_{|[k]\setminus\{t\}}))(s(b)) + \mathrm{H}^0(\varphi_1)(\tilde{b} \to \{t\}; \{i,j\}; b_{|[k]\setminus\{t\}})(s(\tilde{b})).
\end{aligned}
$$

Since $\mathrm{H}^0(\varphi_1)(b \to (\{t\}; \{i,j\}; b_{|[k]\setminus\{t\}}))$ is an isomorphism by the previous lemma, $s(b)$ is determined by $s(\tilde{b})$ which in turn is determined by the values of $s(a)$ with $|\mathrm{im}(a)| = m$ by the induction hypothesis. $\qquad\square$

Since the functor of taking global sections is left-exact, the inclusions

$$
p_*q^*(E_1^{[n]} \otimes \cdots \otimes E_k^{[n]}) = K_k \subset K_1 \subset K_0
$$

induce inclusions

$$
\mathrm{H}^0(X^{[n]}, E_1^{[n]} \otimes \cdots \otimes E_k^{[n]}) \subset \mathrm{H}^0(X^n, K_1)^{\mathfrak{S}_n} \subset \mathrm{H}^0(X^n, K_0)^{\mathfrak{S}_n}.
$$

Furthermore, $\mathrm{H}^0(X^n, K_1) = \ker(\mathrm{H}^0(\varphi_1))$ holds.

Lemma 3.4.3. *Let $n \geq k$. Then the projection*

$$
\mathrm{H}^0(X^n, K_0)^{\mathfrak{S}_n} \cong \bigoplus_{a \in J_0(1)} \mathrm{H}^0(X^n, K_0(a)) \to \mathrm{H}^0(X^n, K_0(1,2,\ldots,k))
$$

induces an isomorphism

$$
\mathrm{H}^0(X^n, K_1)^{\mathfrak{S}_n} \xrightarrow{\cong} \mathrm{H}^0(X^n, K_0(1,2,\ldots,k))
$$

as well as an isomorphism

$$
\mathrm{H}^0(X^{[n]}, E_1 \otimes \cdots \otimes E_k) \xrightarrow{\cong} \mathrm{H}^0(K_0(1,2,\ldots,k)).
$$

Proof. By the previous lemma, the map $\mathrm{H}^0(K_1)^{\mathfrak{S}_n} \to \mathrm{H}^0(K_0(1,2,\ldots,k))$ is injective. Thus it is left to show that for each

$$
x = x_1 \otimes \cdots \otimes x_k \in \mathrm{H}^0(X^n, K_0(1,2,\ldots,k)) = \mathrm{H}^0(E_1) \otimes \cdots \otimes \mathrm{H}^0(E_k)
$$

there exists an $s \in \mathrm{H}^0(X^{[n]}, E_1^{[n]} \otimes \cdots \otimes E_k^{[n]}) \subset \mathrm{H}^0(X^n, K_1)^{\mathfrak{S}_n} \subset \mathrm{H}^0(X^n, K_0)^{\mathfrak{S}_n}$ with $s(1,2,\ldots,k) = x$. We can consider each x_t as a section of $\mathrm{H}^0(X^{[n]}, E_t^{[n]}) \cong \mathrm{H}^0(E_t)$. Then by the last formula of the previous section $s = \Psi(x_1 \otimes \cdots \otimes x_n)$ has the desired property. $\qquad\square$

Theorem 3.4.4. *For $n \geq k$ there is a natural isomorphism*

$$\mathrm{H}^0(X^{[n]}, E_1^{[n]} \otimes \cdots \otimes E_k^{[n]}) \cong \mathrm{H}^0(E_1) \otimes \cdots \otimes \mathrm{H}^0(E_k).$$

Proof. This follows from the lemmas 3.4.3 and 3.2.6. □

3.4.2 Cohomology in degree $2n$

Let for $\ell \in [k]$ be $B_\ell := \mathrm{im}(\varphi_\ell) \subset T_\ell$, i.e. we have exact sequences

$$0 \to K_\ell \to \mathrm{K}_{\ell-1} \to B_\ell \to 0. \tag{1}$$

Since T_ℓ is the push-forward of a sheaf on $\mathbb{D}$, the subsheaf B_ℓ is, too. Since $\dim \mathbb{D} = 2(n-1)$ we have $\mathrm{H}^i(X^n, B_\ell) = 0$ for $i = 2n - 1, 2n$. By the long exact sheaf cohomology sequence associated to (1)

$$\cdots \to \mathrm{H}^{2n-1}(B_\ell) \to \mathrm{H}^{2n}(K_\ell) \to \mathrm{H}^{2n}(K_{\ell-1}) \to \mathrm{H}^{2n}(B_\ell) \to 0$$

we see that $\mathrm{H}^{2n}(K_\ell) = \mathrm{H}^{2n}(K_{\ell-1})$. By induction we get $\mathrm{H}^{2n}(K_\ell) = \mathrm{H}^{2n}(K_0)$. Using corollary 3.1.8 and lemma 3.2.6 this yields the following formula for the cohomology of tensor products of tautological bundles in the maximal degree.

Proposition 3.4.5.

$$\mathrm{H}^{2n}(X^{[n]}, E_1^{[n]} \otimes \cdots \otimes E_k^{[n]}) \cong \mathrm{H}^{2n}(X^n, K_0)^{\mathfrak{S}_n} \cong \bigoplus_{a \in J_0(1)} \bigotimes_{r=1}^{\max a} \mathrm{H}^2(\bigotimes_{t \in a^{-1}(r)} E_t) \otimes S^{n-\max a}\, \mathrm{H}^2(\mathcal{O}_X).$$

3.5 Wedge products in the case of line bundles

Let $E_1 = \cdots = E_k = E$. Then the group $\mathfrak{S}_n$ acts on the tensor product $(E^{[n]})^{\otimes k}$. The wedge product of the tautological bundle is given by the anti-invariants under this action, i.e.

$$\wedge^k E^{[n]} = \left((E^{[n]})^{\otimes k}\right)^{-\mathfrak{S}_k} = \left((E^{[n]})^{\otimes k} \otimes \mathfrak{a}\right)^{\mathfrak{S}_k}.$$

By the results of subsection 1.4.8 we have $\Phi((E^{[n]})^{\otimes n} \otimes \mathfrak{a}) \simeq \Phi((E^{[n]})^{\otimes n}) \otimes \mathfrak{a}$. We also have $\Phi(((E^{[n]})^{\otimes n} \otimes \mathfrak{a})^{\mathfrak{S}_n}) \simeq \Phi((E^{[n]})^{\otimes n} \otimes \mathfrak{a})^{\mathfrak{S}_n}$ since the morphisms p and q are both $\mathfrak{S}_k$-invariant (we consider all varieties equipped with the trivial $\mathfrak{S}_k$-action). Thus, $\Phi(\wedge^k E^{[n]})$ is again concentrated in degree zero and $p_* q^*(\wedge^k E^{[n]}) \cong K_k^{-\mathfrak{S}_k}$. We consider the case $E = L$ with $L \in \mathrm{Pic}\, X$ a line bundle on X and $k \leq n$. Then for every $m \in \mathbb{N}$ the natural $\mathfrak{S}_m$-action on the tensor product $L^{\otimes m}$ given by permutation of the factors equals the identity.

Lemma 3.5.1. *The $\mathfrak{S}_k$-anti-invariants of T_ℓ vanish for $\ell \geq 2$, i.e*

$$T_\ell^{-\mathfrak{S}_k} = (T_\ell \otimes \mathfrak{a}_k)^{\mathfrak{S}_k} = 0 \,.$$

Proof. Let $(M; i, j; a) \in I_\ell$ and t, t' two distinct elements of M. Then $\tau := (t\ t') \in \text{Stab}_{\mathfrak{S}_k}$ acts as the identity on $T_\ell(M; i, j; a)$. Since $\text{sgn}(\tau) = -1$, there are indeed no anti-invariants. $\square$

Corollary 3.5.2. *There is a natural isomorphism $p_* q^*(\wedge^k L^{[n]}) \cong K_1^{-\mathfrak{S}_k}$.*

Proof. This follows by the definition of K_ℓ as the kernel of $\varphi_\ell \colon K_{\ell-1} \to T_\ell$ and the exactness of the functor $(\_ \otimes \mathfrak{a})^{\mathfrak{S}_n}$. $\square$

Lemma 3.5.3. *Let $a \in I_0$ respectively $(M; i, j; a) \in I_1$ such that the map a is not injective. Then the anti-invariants of the corresponding direct summand vanish, which means $K_0(a)^{-\text{Stab}_{\mathfrak{S}_k}(a)} = 0$ respectively $T_1(M; i, j; a)^{-\text{Stab}_{\mathfrak{S}_k}(M; i, j; a)} = 0$.*

Proof. Let t, t' be two distinct numbers in $[k]$ respectively $[k] \setminus M$ with $a(t) = a(t')$. Then $\tau = (t\ t')$ is an element of $\text{Stab}_{\mathfrak{S}_k}(a)$ respectively $\text{Stab}_{\mathfrak{S}_k}(M; i, j; a)$. It acts as the identity which makes the anti-invariants vanish. $\square$

Lemma 3.5.4. *The map $(\varphi_1 \otimes \mathfrak{a}_k)^{\mathfrak{S}_n \times \mathfrak{S}_k} \colon (K_0 \otimes \mathfrak{a})^{\mathfrak{S}_n \times \mathfrak{S}_k} \to (T_1 \otimes \mathfrak{a})^{\mathfrak{S}_n \times \mathfrak{S}_k}$ is zero.*

Proof. We write $\bar{\varphi}_1 = \varphi_1 \otimes \mathfrak{a}$. For $(\{1\}; 1, 2; b) \in J_1(3)$ (see remark 3.2.1) the only possibilities for $b \in \text{Map}([2, k], [n]) \cong [n]^{k-1}$ being injective are

$$b_1 = (1, 2, 3, \ldots, k-1) \quad , \quad b_2 = (1, 3, \ldots, k) \quad , \quad b_3 = (3, \ldots, k+1) \,.$$

Instead of $(\{1\}; 1, 2; b_2)$ we will consider the representative $(\{1\}; 1, 2; b_2')$ with $b_2' = (2, 3, \ldots, k)$ of the same $\mathfrak{S}_n \times \mathfrak{S}_k$-orbit in I_ℓ. The element $((1\ 2), \text{id}_{[k]}) \in \mathfrak{S}_n \times \mathfrak{S}_k$ stabilises $(\{1\}; 1, 2; b_3)$ and acts by -1 on $T_1(\{1\}; 1, 2; b_3)$ which makes the invariants vanish. Thus, it is left to show that $\bar{\varphi}_1(\{1\}; 1, 2; b_1)$ and $\bar{\varphi}_1(\{1\}; 1, 2; b_2')$ vanish on the invariants. For $s \in (K_0 \otimes \mathfrak{a})^{\mathfrak{S}_n \times \mathfrak{S}_k}$ a local section we have indeed

$$\bar{\varphi}_1(s)(\{1\}; 1, 2; b_1) = \big(s(1, 1, 2, \ldots, k-1) - s(2, 1, 2, \ldots, k-1)\big)_{|\Delta_{12}} \overset{3.5.3}{=} (0 - 0)_{|\Delta_{12}} = 0 \,.$$

The element $\sigma = ((1\ 2), (1\ 2))$ acts on $K_0(1, \ldots, k) \otimes \mathfrak{a}_k$ by $\sigma \cdot x = -(1\ 2)_* x$. Since s is a $\mathfrak{S}_n \times \mathfrak{S}_k$ invariant section, $s(1, \ldots, k) = -(1\ 2)_* s(1, \ldots, k)$. Since the restriction of $(1\ 2)_*$ to the diagonal Δ_{12} is the identity, this implies $s(1, \ldots, k)_{|\Delta_{12}} = 0$. Thus,

$$\bar{\varphi}_1(s)(\{1\}; 1, 2; b_2') = \big(s(1, 2, \ldots, k) - s(2, 2, 3 \ldots, k)\big)_{|\Delta_{12}} = 0 - 0 = 0 \,.$$

$\square$

The following two theorems were already proven by Scala in [Sca09a] (theorem 4.2.26 and theorem 5.2.1).

Theorem 3.5.5. *There are natural isomorphism*

$$\mu_*(\wedge^k L^{[n]}) \cong (K_0 \otimes \mathfrak{a})^{\mathfrak{S}_n \times \mathfrak{S}_k} \cong (K_0(1,2,\dots,k) \otimes \mathfrak{a})^{\mathrm{Stab}_{\mathfrak{S}_n \times \mathfrak{S}_k}(1,2,\dots,k)}$$

with $\mathrm{Stab}_{\mathfrak{S}_n \times \mathfrak{S}_k}(1,\dots,k) = \{(\sigma,\sigma) \mid \sigma \in \mathfrak{S}_k\} \times (\mathfrak{S}_{[k+1,n]} \times 1)$.

Proof. This first isomorphism follows from corollary 3.5.2 and lemma 3.5.4 and the second one from lemma 3.5.3. For the description of $\mathrm{Stab}(1,\dots,k)$ see remark 3.2.1. $\qquad\square$

Theorem 3.5.6. *There is a natural isomorphism*

$$\mathrm{H}^*\left(X^{[n]}, \wedge^k L^{[n]}\right) \cong \wedge^k \mathrm{H}^*(L) \otimes S^{n-k} \mathrm{H}^*(A).$$

Proof. The cohomology of $K_0(1,2,\dots,k) \otimes \mathfrak{a}$ on X^n is given by $\mathrm{H}^*(L)^{\otimes k} \otimes \mathrm{H}^*(\mathcal{O}_X)^{\otimes n-k}$. The group $\mathfrak{S}_{[k+1,n]} \leq \mathfrak{S}_n$ acts on $\mathrm{H}^*(\mathcal{O}_X)^{\otimes n-k}$ permuting the tensor factors together with the cohomological sign. Let $\sigma \in \mathfrak{S}_k$. Then $(\sigma,\sigma) \in \mathrm{Stab}_{\mathfrak{S}_n \times \mathfrak{S}_k}(1,2,\dots,k)$ acts on $\mathrm{H}^*(L)^{\otimes k}$ permuting the factors together with the cohomological sign and the sign $\mathrm{sgn}(\sigma)$ since we let $\mathfrak{S}_k$ act by the alternating representation. Now the assertion follows (see section 1.3). $\qquad\square$

3.6 Tensor products of tautological bundles on $X^{[2]}$ and $X^{[n]}_{**}$

3.6.1 Long exact sequences on X^2

We want to enlarge the exact sequences

$$0 \to K_\ell \to K_{\ell-1} \xrightarrow{\varphi_\ell} T_\ell$$

to long exact sequences with a 0 on the right. We will do this first on X^2. Since the pairwise diagonals are disjoint on X^n_{**}, long exact sequences on X^n_{**} can be obtained later from this case. We denote the diagonal as well as its inclusion into X^2 by Δ and its vanishing ideal by $\mathcal{I}$. For a set $M = \{t_1 < \dots < t_s\} \subset [k]$ of cardinality s we will consider the standard representation (see also subsection 1.5.4) $\varrho_M \cong \varrho_s$ of $\mathfrak{S}_M \cong \mathfrak{S}_s$ as the subspace of $\varrho_k \subset \mathbb{C}^k$ with basis

$$\zeta_M^1 := e_{t_1} - e_{t_2}, \, \zeta_M^2 := e_{t_2} - e_{t_3}, \, \dots, \, \zeta_M^{s-1} := e_{t_{s-1}} - e_{t_s}.$$

For $M \subset N$ we denote the inclusion by $\iota_{M \to N} \colon \varrho_M \to \varrho_N$ but will also often omit it in the following. For $\ell = 1,\dots,k$ and $i = 0,\dots,k-\ell$ we set

$$I_\ell^i := \left\{ (M;a) \mid M \subset [k], \, \#M = \ell + i, \, a \colon [k] \setminus M \to [2] \right\}$$

and

$$R_\ell^i := \bigoplus_{(M;a)\in I_\ell^i} \wedge^{\ell-1}\varrho_M(a)$$

where $\varrho_M(a) = \varrho_M$ for every a. We define differentials $d_\ell^i \colon R_\ell^i \to R_\ell^{i+1}$ for $s \in R_\ell^i$ by

$$d_\ell^i(s)(M;a) := \sum_{i\in M} \varepsilon_{i,M}\iota_{M\setminus\{i\}\to M}\left(s(M\setminus\{i\};a,i\mapsto 1) - s(M\setminus\{i\};a,i\mapsto 2)\right).$$

We have indeed defined complexes $R_\ell^\bullet$ for $\ell = 1\ldots,k$ since

$$(d\circ d)(s)(M;a) = \sum_{i\in M} \varepsilon_{i,M}\iota_{M\setminus\{i\}\to M}\big(d(s)(M\setminus\{i\};a,i\mapsto 1) - d(s)(M\setminus\{i\};a,i\mapsto 2)\big)$$

$$= \sum_{i\in M}\sum_{j\in M\setminus\{i\}} \varepsilon_{i,M}\varepsilon_{j,M\setminus\{i\}}\iota_{M\setminus\{i,j\}\to M}\left(\sum_{b\colon \{i,j\}\to[2]} \varepsilon_b s(M\setminus\{i,j\};a\uplus b)\right)$$

vanishes by the fact that $\varepsilon_{i,M}\varepsilon_{j,M\setminus\{i\}} = -\varepsilon_{j,M}\varepsilon_{i,M\setminus\{j\}}$ for all $i,j \in M$. We define a $\mathfrak{S}_k$-action on every R_ℓ^i by setting

$$(\sigma\cdot s)(M;a) := \varepsilon_{\sigma,\sigma^{-1}(M)}\sigma\cdot s(\sigma^{-1}(M);a\circ\sigma).$$

The $\mathfrak{S}_k$-action on the right-hand side is the exterior power of the action on ϱ_k. It maps indeed $\wedge^{\ell-1}\varrho_{\sigma^{-1}(M)}$ to $\wedge^{\ell-1}\varrho_M$. This makes each $R_\ell^\bullet$ into a $\mathfrak{S}_k$-equivariant complex, since for $i \in M$ the term

$$s(\sigma^{-1}(M\setminus\{i\});a\circ\sigma,\sigma^{-1}(i)\mapsto 1) - s(\sigma^{-1}(M\setminus\{i\});a\circ\sigma,\sigma^{-1}(i)\mapsto 2)$$

occurs in $(\sigma\cdot d(s))(M;a)$ with the sign $\varepsilon_{\sigma^{-1}(i),\sigma^{-1}(M)}\cdot \varepsilon_{\sigma,\sigma^{-1}(M)}$ and in $d(\sigma\cdot s)(M;a)$ with the sign $\varepsilon_{i,M}\cdot\varepsilon_{\sigma,\sigma^{-1}(M\setminus\{i\})}$. Both signs are equal by 1.2.1. Note that for $(M;a) \in I_\ell^0$ we have $\wedge^{\ell-1}\varrho_M(a) \cong \mathbb{C}$. We will denote the canonical base vector $\zeta_M^1\wedge\cdots\wedge\zeta_M^{\ell-1}$ by $e_{(M;a)}$. We also define

$$R_\ell^{-1} := \bigoplus_{a\colon [k]\to[2]} \mathbb{C}(a) \quad,\quad \mathbb{C}(a) = \mathbb{C}$$

together with the $\mathfrak{S}_k$-equivariant map $\tilde\varphi_\ell = d_\ell^{-1}\colon R_\ell^{-1} \to R_\ell^0$ given by

$$\tilde\varphi_\ell(s)(M;a) = \left(\sum_{b\colon M\to[2]} \varepsilon_b s(a\uplus b)\right)\cdot e_{(M;a)}.$$

The $\mathfrak{S}_k$-action on R_ℓ^{-1} is given by $(\sigma\cdot s)(a) = s(\sigma^{-1}\circ a)$. We set

$$\tilde R_\ell^\bullet := \left(0\to R_\ell^{-1}\to R_\ell^\bullet\right) = \left(0\to R_\ell^{-1}\to R_\ell^0\to\cdots\to R_\ell^{k-\ell}\to 0\right).$$

We make this complex also $\mathfrak{S}_2$-equivariant by defining the action of $\tau = (1\,2)$ in degree -1 by $(\tau \cdot s)(a) := a(\tau^{-1} \circ a) = a(\tau \circ a)$ and in degree $i \geq 0$ by

$$(\tau \cdot s)(M; a) := \varepsilon^{\ell+i}_{\tau, \{1,2\}} s(M; \tau^{-1} \circ a) = (-1)^{\ell+i} s(M; \tau \circ a).$$

We will sometimes write a k as a left lower index of the occurring objects and morphisms, e.g. $_k R^i_\ell$, if we want to emphasise a chosen value of k.

Proposition 3.6.1. *For every $\ell = 1, \ldots, k$ the complex $\tilde{R}^\bullet_\ell$ is cohomologically concentrated in degree -1, i.e. the sequence*

$$R^{-1}_\ell \to R^0_\ell \to R^1_\ell \to \cdots \to R^{n-\ell}_\ell \to 0$$

is exact.

Proof. We will divide the proof into several lemmas. We will often omit certain indices in the notation, when we think that it will not lead to confusion. For $\ell = 1$ the complex $\tilde{R}^\bullet_1$ is isomorphic to $(\tilde{C}^\bullet)^{\otimes k}[1]$, where $\tilde{C}^\bullet$ is the complex concentrated in degree 0 and 1 given by

$$0 \to \mathbb{C} \oplus \mathbb{C} \to \mathbb{C} \to 0 \quad , \quad \begin{pmatrix} a \\ b \end{pmatrix} \mapsto a - b.$$

Since the complex $\tilde{C}^\bullet$ has only cohomology in degree zero and the tensor product is taken over the field $\mathbb{C}$, it follows that $\tilde{R}^\bullet_1$ is indeed cohomologically concentrated in degree -1. We go on by induction over ℓ assuming that the proposition is true for all values smaller than ℓ.

Lemma 3.6.2. *Let $t \in R^0_\ell$. Then $d^0_\ell(t) = 0$ if and only if for every $(N; a) \in I^1_\ell$ and every pair $i, j \in N$ the following holds:*

$$t(N \setminus \{i\}; a, i \mapsto 1) - t(N \setminus \{i\}; a, i \mapsto 2) = t(N \setminus \{j\}; a, j \mapsto 1) - t(N \setminus \{j\}; a, j \mapsto 2).$$

Here for $(M; b) \in I^0_\ell$ we use the notation $t(M; b) = t(M; b) \cdot e_{(M;b)}$, i.e. we denote by $t(M; b)$ also its preimage under the canonical isomorphism $\mathbb{C} \cong \wedge^{\ell-1} \varrho_M$.

Proof. Let $N = \{n_1 < \cdots < n_{\ell+1}\} \subset [k]$. The above formula holds for every pair $i, j \in N$ if and only if it holds for every pair of neighbours. Thus we may assume that $i = n_h$ and $j = n_{h+1}$ with $h \in [\ell]$. The wedge product $\wedge^{\ell-1} \varrho_{N \setminus \{n_h\}}$ is spanned by the vector

$$e_{N \setminus \{n_h\}} = \zeta^1_{N \setminus \{n_h\}} \wedge \cdots \wedge \zeta^{\ell-1}_{N \setminus \{n_h\}} = \begin{cases} \zeta^2_N \wedge \cdots \wedge \zeta^\ell_N & \text{for } h = 1, \\ \zeta^1_N \wedge \cdots \wedge (\zeta^{h-1}_N + \zeta^h_N) \wedge \cdots \wedge \zeta^\ell_N & \text{else.} \end{cases}$$

Thus, for $t \in R_\ell^0$ the coefficient of $\zeta_N^1 \wedge \cdots \wedge \widehat{\zeta_N^h} \wedge \cdots \wedge \zeta_N^\ell$ of $d(t)(N,a) \in \wedge^{\ell-1}\varrho_N$ equals

$$\varepsilon_{n_h,N}\big(t(N \setminus \{n_h\}; a, n_h \mapsto 1) - t(N \setminus \{n_h\}; a, n_h \mapsto 2)\big)$$
$$+\varepsilon_{n_{h+1},N}\big(t(N \setminus \{n_{h+1}\}; a, n_{h+1} \mapsto 1) - t(N \setminus \{n_{h+1}\}; a, n_{h+1} \mapsto 2)\big)$$

which proves the lemma. $\qquad\square$

The inclusion $\mathrm{im}(\tilde\varphi) \subset \ker(d_\ell^0)$ follows since for $s \in R_\ell^{-1}$ and $t = \tilde\varphi(s)$ both sides of the equation in the above lemma equal

$$\sum_{b:\, N \to [2]} \varepsilon_b \cdot s(a \uplus b).$$

We will actually show a bit more than $\mathrm{im}(\tilde\varphi) \supset \ker(d_\ell^0)$ in the next lemma. We decompose R_ℓ^{-1} into $U_\ell \oplus S_\ell$ with

$$U_\ell = \{s \in R_\ell^{-1} \mid s(a) = 0 \,\forall a : |a^{-1}(\{2\})| \leq \ell - 1\} = \langle e_a \mid a^{-1}(\{2\}) \geq \ell \rangle$$
$$S_\ell = \{s \in R_\ell^{-1} \mid s(a) = 0 \,\forall a : |a^{-1}(\{2\})| \geq \ell\} = \langle e_a \mid a^{-1}(\{2\}) \leq \ell - 1 \rangle.$$

Lemma 3.6.3. *The map $\tilde\varphi_{\ell|U_\ell} : U_\ell \to \ker(d_\ell^0)$ is an isomorphism.*

Proof. We first show the injectivity. Let $s \in U$ with $\tilde\varphi(s) = 0$. We show that $s(a)$ is zero for every a by induction over $\alpha = |a^{-1}(\{2\})|$. For $\alpha = \ell$ we set $M = a^{-1}(\{2\})$ and get

$$0 = \tilde\varphi(s)(M; \underline{1}) = s(a).$$

We may now assume that $s(b) = 0$ for every $b : [k] \to [2]$ with $b^{-1}(\{2\}) < \alpha$. Then for $M \subset a^{-1}(\{2\})$ with $|M| = \ell$ we obtain

$$0 = \tilde\varphi(s)(M; a_{|[k]\setminus M}) = s(a).$$

For the surjectivity we precede by induction on k. For $k = \ell$ the map $\tilde\varphi$ sends the basis vector $e_{\underline{2}}$ of the one-dimensional vector space U with a factor of $(-1)^\ell$ to the basis vector $e_{[\ell]}$ of the one-dimensional vector space T_ℓ^0. Let now $k > \ell$ be arbitrary. We set

$$V_0 = \{a : [k] \to [2] \mid a(k) = 2, \#a^{-1}(\{2\}) = \ell\},$$
$$V_1 = \{a : [k] \to [2] \mid a(k) = 1, \#a^{-1}(\{2\}) \geq \ell\},$$
$$V_2 = \{a : [k] \to [2] \mid a(k) = 2, \#a^{-1}(\{2\}) \geq \ell + 1\},$$
$$W_0 = \{(M,a) \in {}_kI_\ell^0 \mid k \in M, a = \underline{1}\}, \ W_1 = \{(M,a) \in {}_kI_\ell^0 \mid k \notin M, a(k) = 1\},$$
$$W_2 = \{(M,a) \in {}_kI_\ell^0 \mid k \notin M, a(k) = 2\}, \ W_3 = \{(M,a) \in {}_kI_\ell^0 \mid k \in M, a \neq \underline{1}\},$$

We define subspaces of U respectively T_ℓ^0 by

$$\langle V_i \rangle = \langle e_a \mid a \in V_i \rangle \quad , \quad \langle W_i \rangle = \langle e_{(M;a)} \mid (M;a) \in W_i \rangle$$

which yields $U = \langle V_0 \rangle \oplus \langle V_1 \rangle \oplus \langle V_2 \rangle$ and $T_\ell^0 = \langle W_0 \rangle \oplus \langle W_1 \rangle \oplus \langle W_2 \rangle \oplus \langle W_3 \rangle$. We denote by $\tilde{\varphi}(i,j)$ the component of $\tilde{\varphi}_\ell$ given by the composition

$$\langle V_i \rangle \to U \xrightarrow{\tilde{\varphi}} T_\ell^0 \to \langle W_j \rangle \,.$$

Let now $t \in \ker({}_k d_\ell^0)$. We define $s_0 \in \langle V_0 \rangle$ by $s_0(a) = t(a^{-1}(\{2\}); \underline{1})$ for $a \in V_0$. This yields $\tilde{\varphi}(0,0)(s_0) = t_{|\langle W_0 \rangle}$. We set $\tilde{t} := t - \tilde{\varphi}(s_0)$. The components $\tilde{\varphi}(1,0)$, $\tilde{\varphi}(2,0)$, $\tilde{\varphi}(1,2)$, and $\tilde{\varphi}(2,1)$ are all zero. There are canonical bijections

$$V_1 \xrightarrow{\cong} \{a \colon [k-1] \to [2] \mid \#a^{-1}(\{2\}) \geq \ell\} \xleftarrow{\cong} V_2 \,,$$
$$W_1 \xrightarrow{\cong} \{(M,a) \mid M \subset [k-1], \#M = \ell, a \colon [k-1] \setminus M \to [2]\} \xleftarrow{\cong} W_2$$

given by dropping $a(k)$. They induce linear isomorphisms $\langle V_1 \rangle \cong {}_{k-1}U_\ell \cong \langle V_2 \rangle$ as well as $\langle W_1 \rangle \cong {}_{k-1}T_\ell^0 \cong \langle W_2 \rangle$ under which the linear maps $\tilde{\varphi}(1,1)$ and $\tilde{\varphi}(2,2)$ both correspond to ${}_{k-1}\tilde{\varphi}$. Thus, by induction there are $s_1 \in \langle V_1 \rangle$ and $s_2 \in \langle V_2 \rangle$ such that $\tilde{\varphi}(1,1)(s_1) = \tilde{t}_{|\langle W_1 \rangle}$ and $\tilde{\varphi}(2,2)(s_2) = \tilde{t}_{|\langle W_2 \rangle}$. Defining $s \in {}_kU$ by $s_{|\langle V_i \rangle} = s_i$ for $i = 0,1,2$ we get

$$\tilde{\varphi}(s)_{|\langle W_0 \rangle \oplus \langle W_1 \rangle \oplus \langle W_2 \rangle} = t_{|\langle W_0 \rangle \oplus \langle W_1 \rangle \oplus \langle W_2 \rangle} \,.$$

It is left to show that the equation also holds on $\langle W_3 \rangle$. For this we show that for every $x \in \ker d_\ell^0$ with $x_{|\langle W_0 \rangle \oplus \langle W_1 \rangle \oplus \langle W_2 \rangle} = 0$ also $x(M;a) = 0$ for every $(M,a) \in W_3$ holds. We use induction over $\alpha = |a^{-1}(\{2\})|$. For $\alpha = 0$ the tuple $(M;a)$ is an element of W_0. Hence, $x(M;a) = 0$ holds. We now assume that $x(M;b) = 0$ for every tuple $(M;b) \in W_3$ with $|b^{-1}(\{2\})| < \alpha = |a^{-1}(\{2\})|$ holds and choose an $i \in [k] \setminus M$ with $a(i) = 2$. Applying lemma 3.6.2 to $N = M \cup \{i\}$ we get

$$x(M;a) = x(M; a_{|[k]\setminus N}, i \mapsto 1) - x(N \setminus \{k\}; a_{|[k]\setminus N}, k \mapsto 1) + x(N \setminus \{k\}; a_{|[k]\setminus N}, k \mapsto 2) \,.$$

The first term on the right hand side is zero by induction hypothesis and the second and third are zero since they are coefficients of s in W_1 respectively W_2. $\qquad\square$

Remark 3.6.4. Counting the cardinality of the base we get

$$\dim({}_kU_\ell) = \sum_{j=\ell}^{k} \binom{k}{j} \,.$$

On the other hand we have

$$\dim(_k R_\ell^i) = 2^{k-\ell-i} \binom{k}{\ell+i} \binom{\ell+i-1}{\ell-1}$$

and thus also

$$\chi(_k R_\ell^\bullet) = \sum_{i=0}^{k-\ell} (-1)^i 2^{k-\ell-i} \binom{k}{\ell+i} \binom{\ell+i-1}{\ell-1} \overset{1.2.2}{=} \sum_{j=\ell}^{k} \binom{k}{j}.$$

It follows by the previous lemma that if the complex $R_\ell^\bullet$ is exact in all but one positive degree it is exact in every positive degree.

Lemma 3.6.5. *Let $k \geq \ell + 2$ and $2 \leq j \leq k - \ell$. Then $\mathcal{H}^j(R_\ell^\bullet) = 0$, i.e. the sequence*

$$R_\ell^{j-1} \to R_\ell^j \to R_\ell^{j+1}$$

is exact (with $R_\ell^{j+1} = 0$ in the case $j = k - \ell$).

Proof. The term $T_\ell^{k-\ell} = \wedge^{\ell-1} \varrho_{[k]}$ is an irreducible $\mathfrak{S}_k$-representation (see [FH91, Proposition 3.12]). Since $d_\ell^{k-\ell-1}$ is non-zero and $\mathfrak{S}_k$-equivariant, it follows that it is surjective. This proves the case $j = k - \ell$. For general $j \in [2, k-\ell]$ we use induction over k. For $k = \ell + 2$ only the case $j = k - \ell$ occurs which is already proven. So now let $k \geq \ell + 3$, $2 \leq j \leq k - \ell - 1$, and $t \in \ker d_\ell^j$. For $i = 0, \ldots, k - \ell$ we decompose R_ℓ^i into the direct summands

$$R_\ell^i(0) = \bigoplus_{(M,a): \, k \in M} \wedge^{\ell-1} \varrho_M(a), \quad R_\ell^i(1) = \bigoplus_{(M,a): \, k \notin M, \, a(k)=1} \wedge^{\ell-1} \varrho_M(a)$$

$$R_\ell^i(2) = \bigoplus_{(M,a): \, k \notin M, \, a(k)=2} \wedge^{\ell-1} \varrho_M(a)$$

and denote the components of the differential by $d(u,v) \colon R_\ell^i(u) \to R_\ell^{i+1}(v)$. Then $d(0,1)$, $d(0,2)$, $d(1,2)$ and $d(2,1)$ all vanish. It follows that $t_1 \in \ker d(1,1)$ and $t_2 \in \ker d(2,2)$ where t_u is the component of t in $R_\ell^j(u)$. By dropping $a(k)$ we get isomorphisms

$$_k R_\ell^i(1) \cong {}_{k-1} R_\ell^i \cong {}_k R_\ell^i(2)$$

under which $_k d_\ell^i(1,1)$ as well as $_k d_\ell^i(2,2)$ coincide with $_{k-1} d_\ell^i$. Thus, by induction there exist $s_u \in {}_k R_\ell^{j-1}(u)$ for $u = 1, 2$ such that $d(u,u)(s_u) = t_u$. It remains to find a $s_0 \in R_\ell^{j-1}(0)$ such that

$$d(s_0) = d(0,0)(s_0) = \hat{t} := t - d(s_1 + s_2).$$

74

For $M \subset [k-1]$ we have $\varrho_{M \cup \{k\}} = \varrho_M \oplus \langle \zeta_{M \cup \{k\}, \max} \rangle$, where $\max = |M|$, i.e.

$$\zeta_{M \cup \{k\}, \max} = \zeta_{M \cup \{k\}, |M|} = e_{\max(M)} - e_k \, .$$

Using this we can decompose the $R_\ell^i(0)$ further into the two direct summands

$$R_\ell^i(3) = \bigoplus_{(M,a): \, k \in M} \wedge^{\ell-1} \varrho_{M \setminus \{k\}}(a) \, , \; R_\ell^i(4) = \bigoplus_{(M,a): \, k \in M} \wedge^{\ell-2} \varrho_{M \setminus \{k\}} \otimes \langle \zeta_{M, \max} \rangle(a) \, .$$

The differential $d(3,4)$ vanishes. Thus, $d(4,4)(\hat{t}_4) = 0$. Furthermore we have the following isomorphism of sequences

$$
\begin{array}{ccccc}
{}_{k-1}R_{\ell-1}^{j-1} & \xrightarrow{{}_{k-1}d_{\ell-1}^{j-1}} & {}_{k-1}R_{\ell-1}^{j} & \xrightarrow{{}_{k-1}d_{\ell-1}^{j}} & {}_{k-1}R_{\ell-1}^{j+1} \\
\cong \downarrow & & \cong \downarrow & & \cong \downarrow \\
{}_{k}R_{\ell}^{j-1}(4) & \xrightarrow{{}_{k}d_{\ell}^{j-1}(4,4)} & {}_{k}R_{\ell}^{j}(4) & \xrightarrow{{}_{k}d_{\ell}^{j}(4,4)} & {}_{k}R_{\ell}^{j+1}(4)
\end{array}
$$

induced by the maps ${}_{k-1}I_{\ell-1}^i \to {}_kI_\ell^i$ given by $(L;a) \mapsto (L \cup \{k\};a)$ and the isomorphisms $\wedge^{\ell-2}\varrho_{M \setminus \{k\}} \cong \wedge^{\ell-2}\varrho_{M \setminus \{k\}} \otimes \langle \zeta_{M, \max} \rangle$. By the induction hypothesis for ℓ (saying that proposition 3.6.6 is true for $\ell - 1$) the upper sequence is exact. Thus, there is a $s_4 \in R_\ell^{j-1}(4)$ such that $d(4,4)(s_4) = \hat{t}_4$. We set $\tilde{t}_3 := \hat{t}_3 - d(4,3)(s_4)$. There is also a canonical isomorphism of the exact sequences

$$
\begin{array}{ccccc}
{}_{k-1}R_{\ell}^{j-2} & \xrightarrow{{}_{k-1}d_{\ell}^{j-2}} & {}_{k-1}R_{\ell}^{j-1} & \xrightarrow{{}_{k-1}d_{\ell}^{j-1}} & {}_{k-1}R_{\ell}^{j} \\
\cong \downarrow & & \cong \downarrow & & \cong \downarrow \\
{}_{k}R_{\ell}^{j-1}(3) & \xrightarrow{{}_{k}d_{\ell}^{j-1}(3,3)} & {}_{k}R_{\ell}^{j}(3) & \xrightarrow{{}_{k}d_{\ell}^{j}(3,3)} & {}_{k}R_{\ell}^{j+1}(3)
\end{array}
$$

induced by the maps ${}_{k-1}I_\ell^{i-1} \to {}_kI_\ell^i$ which are again given by $(L;a) \mapsto (L \cup \{k\};a)$. By induction over k the upper sequence is exact (in the case that $j = 2$ we have to use remark 3.6.4). This yields a $s_3 \in R_\ell^{j-1}(3)$ such that $d(3,3)(s_3) = \tilde{t}_3$ holds. In summary we have found a $s = (s_1, s_2, s_3, s_4) \in R_\ell^{j-1}$ with $d(s) = t$. $\qquad\Box$

As mentioned in remark 3.6.4 the previous lemma finishes the proof of proposition 3.6.1. $\qquad\Box$

For $\ell = 1, \dots, k$ we set

$$H_\ell := \Delta_* \left(S^{\ell-1}\Omega_X \otimes E_1 \otimes \cdots \otimes E_k \right) = \left(S^{\ell-1}\Omega_X \otimes E_1 \otimes \cdots \otimes E_k \right)_{12} \, .$$

Then H_ℓ equals $T_\ell(M; 1, 2; a)$ (see section 3.1.1) for every tuple $(M; a) \in I_\ell^0 \cong {}_2I_\ell$. We define the complexes $T_\ell^\bullet := H_\ell \otimes_{\mathbb{C}} R_\ell^\bullet$ and $\tilde{T}_\ell^\bullet = H_\ell \otimes_{\mathbb{C}} \tilde{R}_\ell^\bullet$. We denote again the differentials by d_ℓ^i and $d_\ell^{-1} = \tilde{\varphi}_\ell$. Then for every $\ell = 1, \ldots, k$ there are isomorphisms $T_\ell \cong T_\ell^0$. We will always consider the isomorphism $T_\ell \cong T_\ell^0$ induced by the canonical isomorphisms $\mathbb{C} \cong \wedge^{\ell-1} \varrho_M(a)$ for $(M; a) \in I_\ell^0$ given by $1 \mapsto e_{(M;a)}$. We will denote the composition of $\varphi_\ell \colon K_{\ell-1} \to T_\ell$ with this isomorphism again by $\varphi_\ell \colon K_{\ell-1} \to T_\ell^0$. If $E_1 = \cdots = E_k$, the complex $\tilde{T}_\ell^\bullet$ carries a canonical $\mathfrak{S}_k$-linearization given by applying the action of $\mathfrak{S}_k$ on $\tilde{R}_\ell^\bullet$ and permuting the tensor factors of H_ℓ. We also define a $\mathfrak{S}_2$-action on $\tilde{T}_\ell^\bullet$ by using the $\mathfrak{S}_2$-action on $\tilde{R}_\ell^\bullet$ as well as the natural action on H_ℓ which means that τ acts on H_ℓ by $(-1)^{\ell-1}$ because of the factor $(S^{\ell-1}\Omega_X)_{12} = S^{\ell-1}N_\Delta^\vee$. The isomorphism $T_\ell \cong T_\ell^0$ is $\mathfrak{S}_k$-equivariant since

$$\sigma \cdot e_{(\sigma^{-1}(M), \sigma^{-1} \circ a)} = \sigma \cdot (\zeta_{\sigma^{-1}(M)}^1 \wedge \cdots \wedge \zeta_{\sigma^{-1}(M)}^{\ell-1}) = \varepsilon_{\sigma, \sigma^{-1}(M)} \cdot \zeta_M^1 \wedge \cdots \wedge \zeta_M^{\ell-1} = \varepsilon_{\sigma, \sigma^{-1}(M)} \cdot e_{(M;a)} \, .$$

It is also $\mathfrak{S}_2$-equivariant. Hence, $\varphi_\ell \colon K_{\ell-1} \to T_\ell^0$ is again $\mathfrak{S}_2$-equivariant and, if all the E_i are equal, also $\mathfrak{S}_k$-invariant.

Proposition 3.6.6. *For every $\ell = 1, \ldots, k$ the sequences*

$$0 \to K_\ell \to K_{\ell-1} \xrightarrow{\varphi_\ell} T_\ell^0 \to T_\ell^1 \to \cdots \to T_\ell^{k-\ell} \to 0$$

are exact.

Proof. With the same arguments as in the proof of proposition 3.1.5 we may assume that $E_1 = \cdots = E_k = \mathcal{O}_X$. Thus we have

$$H_\ell = \Delta_* \left(S^{\ell-1}\Omega_X \otimes E_1 \otimes \cdots \otimes E_k \right) = \mathcal{I}^{\ell-1}/\mathcal{I}^\ell \, .$$

A section $s \in K_0$ with arbitrary values $s(a_1, \ldots, a_k) \in \mathcal{I}^{\ell-1} \subset \mathcal{O}_{X^2}$ is already in the kernel of all the maps $\varphi_1, \ldots, \varphi_{\ell-1}$, i.e. is a section of $K_{\ell-1}$. Thus, the image of $\varphi_\ell \colon K_{\ell-1} \to T_\ell^0$ equals the image of $\tilde{\varphi}_\ell \colon T_\ell^{-1} \to T_\ell^0$. Hence, it suffices to show the exactness of

$$T_\ell^{-1} \xrightarrow{\tilde{\varphi}_\ell} T_\ell^0 \to T_\ell^1 \to \cdots \to T_\ell^{k-\ell} \to 0 \, .$$

Since the tensor product over $\mathbb{C}$ is exact, this follows from proposition 3.6.1. $\square$

Remark 3.6.7. Using lemma 3.6.3 and the exactness of the tensor product over $\mathbb{C}$ we get also a description of the K_ℓ as kernels of the short exact sequences

$$0 \to K_\ell \to K_{\ell-1} \to H_\ell \otimes_{\mathbb{C}} U_\ell \to 0 \, .$$

However, as long as we do not have an explicit description of the inverse $\ker(d_\ell^0) \to U_\ell$ of $\tilde{\varphi}_{\ell|U_\ell}$, this description of $p_*q^*(E_1^{[n]} \otimes \cdots \otimes E_k^{[n]})$ is not very useful since we neither know how the maps $K_\ell \to H_\ell \otimes_C U_\ell$ look like nor how we have to define the $\mathfrak{S}_2$-linearization on $H_\ell \otimes_\mathbb{C} U_\ell$.

3.6.2 The invariants on $S^2 X$

For $\ell = 1, \ldots, k$ we define the complex $\hat{R}_\ell^\bullet$ of $\mathfrak{S}_2 \times \mathfrak{S}_k$-representations by setting $\hat{R}_\ell^\bullet = R_\ell^\bullet$ as $\mathfrak{S}_k$-representations and defining the $\mathfrak{S}_2$-action by letting $\tau = (1\ 2)$ act on $\hat{R}_\ell^i$ by

$$(\tau \cdot s)(M; a) = (-1)^{\ell+i+\ell-1} s(M; \tau^{-1} \circ a) = (-1)^{i-1} s(M; \tau \circ a).$$

In this way we have $T_\ell^\bullet = H_\ell \otimes_C \hat{R}_\ell^\bullet$ as $\mathfrak{S}_2 \times \mathfrak{S}_k$-equivariant complexes, when considering H_ℓ equipped with the trivial action. The action of τ on the index set I_ℓ^i of the direct sum $\hat{R}_\ell^i$ is given by $\tau \cdot (M; a) = (M; \tau \circ a)$. We define for $\ell \leq k$ the number $N(k, \ell)$ by

$$N(k, \ell) = \frac{1}{2} \left(\sum_{j=\ell}^{k} \binom{k}{j} - \binom{k-1}{\ell-1} \right).$$

Note that $N(k, k) = 0$.

Lemma 3.6.8. *The following holds for two natural numbers $\ell \leq k$:*

$$\dim \left(\ker({}_k \hat{d}_\ell^0)^{\mathfrak{S}_2} \right) = \dim \left(\mathcal{H}^0({}_k \hat{R}_\ell^\bullet)^{\mathfrak{S}_2} \right) = \chi \left(({}_k \hat{R}_\ell^\bullet)^{\mathfrak{S}_2} \right) = N(k, \ell).$$

Proof. The second equality follows by proposition 3.6.1 and the fact that taking invariants is exact. The only non-trivial element $\tau = (1\ 2)$ of $\mathfrak{S}_2$ acts freely on I_ℓ^i for $0 \leq i < k - \ell$. Thus, by Danila's lemma $\dim((R_\ell^i)^{\mathfrak{S}_2}) = 1/2 \dim(R_\ell^i)$ for $i < k - \ell$. Furthermore, τ acts by $(-1)^{k+\ell-1} = (-1)^{k-\ell-1}$ on $\hat{R}_\ell^{k-\ell}$. Hence we have $(\hat{R}_\ell^{k-\ell})^{\mathfrak{S}_2} = \hat{R}_\ell^{k-\ell}$ if $k - \ell$ is odd and $(\hat{R}_\ell^{k-\ell})^{\mathfrak{S}_2} = 0$ if $k - \ell$ is even. The assertion follows using remark 3.6.4. $\square$

Proposition 3.6.9. *On $S^2 X$ there are for $\ell = 1, \ldots, k$ exact sequences*

$$0 \to K_\ell^{\mathfrak{S}_2} \to K_{\ell-1}^{\mathfrak{S}_2} \to (\pi_\star H_\ell)^{\oplus N(k,\ell)} \to 0.$$

In particular $K_k^{\mathfrak{S}_2} = K_{k-1}^{\mathfrak{S}_2}$.

Proof. By proposition 3.6.6 we have $\operatorname{im}(\varphi_\ell) = \mathcal{H}^0(T_\ell^\bullet)$. Since the tensor product over $\mathbb{C}$ as well as taking $\mathfrak{S}_2$-invariants are exact functors and we consider H_ℓ equipped with the trivial $\mathfrak{S}_2$-action,

$$\operatorname{im}(\varphi_\ell)^{\mathfrak{S}_2} = \mathcal{H}^0(T_\ell^\bullet)^{\mathfrak{S}_2} = H_\ell \otimes_\mathbb{C} \mathcal{H}^0(\hat{R}_\ell^\bullet)^{\mathfrak{S}_2} = (\pi_\star H_\ell)^{\oplus N(k,\ell)}$$

follows by the previous lemma. $\square$

The composition $\delta := \pi \circ \Delta$ is the diagonal embedding of X into $S^2 X$.

Corollary 3.6.10. *For $E_1, \ldots, E_k$ locally free sheaves on X, there is in the Grothendieck group $\mathrm{K}(S^2 X)$ the equality*

$$\mu_! \left[(E_1^{[2]} \otimes \cdots \otimes E_k^{[2]}) \right] = \left[K_0^{\mathfrak{S}_2} \right] - \sum_{\ell=1}^{k} N(k, \ell) \delta_! \left(\left[S^{\ell-1} \Omega_X \right] \cdot [E_1] \cdots [E_k] \right).$$

Proof. Since $R\mu_*(E_1^{[2]} \otimes \cdots \otimes E_k^{[2]})$ is cohomologically concentrated in degree zero (see corollary 3.8.3), we have

$$\mu_! \left[(E_1^{[2]} \otimes \cdots \otimes E_k^{[2]}) \right] = \left[\mu_*(E_1^{[2]} \otimes \cdots \otimes E_k^{[2]}) \right].$$

Now the formula follows by $K_k^{\mathfrak{S}_2} = \mu_*(E_1^{[2]} \otimes \cdots \otimes E_k^{[2]})$ (see corollary 3.1.8) and the previous proposition. □

We can identify the set $_2J_0(1)$ of representatives of the $\mathfrak{S}_2$-orbits of $_2I_0$ (see remark 3.2.1 (i)) with the set of partitions $P = \{P_1, P_2\}$ of the set $[k]$ of length at most two (we denote the case of length one by $P = \{[k], \emptyset\}$). For this we assume that $P_1 \prec P_2$ in the total order defined in subsection 3.2.1. Then we identify P with the map $a \colon [k] \to [2]$ with $a^{-1}(1) = P_1$ and $a^{-1}(2) = P_2$. We get the following as a special case of lemma 3.2.2.

Lemma 3.6.11. *On $S^2 X$ there is the following isomorphism*

$$K_0^{\mathfrak{S}_2} \cong \bigoplus_{P = \{P_1, P_2\} \in J_0(1)} \pi_* K_0(P) \quad , \quad K_0(P) = \left(\bigotimes_{t \in P_1} E_t \right) \boxtimes \left(\bigotimes_{t \in P_2} E_t \right).$$

3.6.3 Cohomology on $X^{[2]}$

Lemma 3.6.12. *For every $P = \{P_1, P_2\} \in J_0(1)$ there is a canonical isomorphism*

$$\mathrm{H}^*(X^2, K_0(P)) \cong \mathrm{H}^*(\otimes_{t \in P_1} E_t) \otimes \mathrm{H}^*(\otimes_{t \in P_2} E_t).$$

This yields

$$\mathrm{H}^*(S^2 X, K_0^{\mathfrak{S}_2}) = \bigoplus_{\{P_1, P_2\} \in J_0} \mathrm{H}^* \left(\bigotimes_{t \in P_1} E_t \right) \otimes \mathrm{H}^* \left(\bigotimes_{t \in P_1} E_t \right).$$

Proof. This follows by lemma 3.6.11, the Künneth formula, and the fact that π_* is exact. □

Lemma 3.6.13. *Let $E_1, \ldots, E_k$ be locally free sheaves on X and $\ell = 1, \ldots, k$. Then*

$$\mathrm{H}^*(X^2, H_\ell) = \mathrm{H}^*(S^2 X, \pi_* H_\ell) = \mathrm{H}^*(X, S^{\ell-1} \Omega_X \otimes E_1 \otimes \cdots \otimes E_k).$$

Proof. The first equation is because of π being exact and the second one because of $\Delta = \iota_{12}$ being exact. Alternatively, this and the previous lemma can be obtained as special cases of lemma 3.2.6. □

Lemma 3.6.14. *For every* $\ell = 1, \ldots, k-1$ *and every* $i \geq 0$ *we have*

$$\mathrm{H}^i\left(X^2, \ker(d_\ell^0)\right)^{\mathfrak{S}_2} = \mathrm{H}^i\left(S^2X, \ker(d_\ell^0)^{\mathfrak{S}_2}\right) = \mathrm{H}^i\left(S^2X, \ker((d_\ell^0)^{\mathfrak{S}_2})\right)$$
$$= \ker\left(\mathrm{H}^i(S^2X, (d_\ell^0)^{\mathfrak{S}_2})\right) = \ker\left(\mathrm{H}^i(X^2, d_\ell^0)^{\mathfrak{S}_2}\right) = \ker\left(\mathrm{H}^i(X^2, d_\ell^0)\right)^{\mathfrak{S}_2}$$

where $d_\ell^0 \colon T_\ell^0 \to T_\ell^1$ *is the differential.*

Proof. All but the third equation follow directly by the fact that taking the invariants is an exact functor. We have $(T_\ell^\bullet)^{\mathfrak{S}_2} = H_\ell \otimes (\hat{R}_\ell^\bullet)^{\mathfrak{S}_2}$. In the category of vector spaces every exact sequence, and thus in particular

$$0 \to \ker(\hat{d}_\ell^0)^{\mathfrak{S}_2} \to (\hat{R}_\ell^0)^{\mathfrak{S}_2} \to \mathrm{im}(\hat{d}_\ell^0)^{\mathfrak{S}_2} \to 0\,,$$

splits. Since the functor $H_\ell \otimes_{\mathbb{C}} \_$ is exact this makes also the sequence

$$0 \to \ker(d_\ell^0)^{\mathfrak{S}_2} \to (T_\ell^0)^{\mathfrak{S}_2} \to \mathrm{im}(d_\ell^0)^{\mathfrak{S}_2} \to 0$$

of sheaves on S^2X split. Hence, the associated long exact sequence in cohomology decomposes into the short exact sequences

$$0 \to \mathrm{H}^i\left(\ker(d_\ell^0)^{\mathfrak{S}_2}\right) \to \mathrm{H}^i\left((T_\ell^0)^{\mathfrak{S}_2}\right) \to \mathrm{H}^i\left(\mathrm{im}(d_\ell^0)^{\mathfrak{S}_2}\right) \to 0$$

which gives $\mathrm{H}^i(\ker(d_\ell^0)^{\mathfrak{S}_2}) = \ker\bigl(\mathrm{H}^i((T_\ell^0)^{\mathfrak{S}_2}) \to \mathrm{H}^i(\mathrm{im}(d_\ell^0)^{\mathfrak{S}_2})\bigr)$. Now by the same arguments as before the sequence

$$0 \to \mathrm{H}^i\left(\mathrm{im}(d_\ell^0)^{\mathfrak{S}_2}\right) \to \mathrm{H}^i\left((T_\ell^1)^{\mathfrak{S}_2}\right) \to \mathrm{H}^i\left(\mathrm{im}(d_\ell^1)^{\mathfrak{S}_2}\right) \to 0$$

splits. Thus, the map $\mathrm{H}^i(\mathrm{im}(d_\ell^0)^{\mathfrak{S}_2}) \to \mathrm{H}^i((T_\ell^1)^{\mathfrak{S}_2})$ is injective and

$$\ker\left(\mathrm{H}^i((T_\ell^0)^{\mathfrak{S}_2}) \to \mathrm{H}^i(\mathrm{im}(d_\ell^0)^{\mathfrak{S}_2})\right) = \ker\left(\mathrm{H}^i((T_\ell^0)^{\mathfrak{S}_2}) \xrightarrow{(d_\ell^0)^{\mathfrak{S}_2}} \mathrm{H}^i((T_\ell^1)^{\mathfrak{S}_2})\right)\,.$$

□

We set $D_\ell^i = \ker(\mathrm{H}^i(S^2X, (d_\ell^0)^{\mathfrak{S}_2}))$. It is abstractly isomorphic to

$$\mathrm{H}^i(S^{\ell-1}\Omega_X \otimes E_1 \otimes \cdots \otimes E_k)^{\oplus N(k,\ell)}\,.$$

By what is done so far, we get the following description of $\mathrm{H}^*(X^{[2]}, E_1^{[2]} \otimes \cdots \otimes E_k^{[2]})$ in terms of sheaf cohomology on X for locally free sheaves $E_1, \ldots, E_k$.

79

Proposition 3.6.15. *There are for $\ell = 1, \ldots, k-1$ long exact sequences*

$$\cdots \to D_\ell^{i-1} \to \mathrm{H}^i(S^2 X, K_\ell^{\mathfrak{S}_2}) \to \mathrm{H}^i(S^2 X, K_{\ell-1}^{\mathfrak{S}_2}) \to D_\ell^i \to \mathrm{H}^{i+1}(S^2 X, K_\ell^{\mathfrak{S}_2}) \to \cdots$$

with $\mathrm{H}^*(S^2 X, K_0^{\mathfrak{S}_2}) = \oplus_{\{P_1, P_2\} \in J_0(1)} \mathrm{H}^*(\otimes_{t \in P_1} E_t) \otimes \mathrm{H}^*(\otimes_{t \in P_1} E_t)$ *and*

$$\mathrm{H}^*(S^2 X, K_{k-1}^{\mathfrak{S}_2}) = \mathrm{H}^*(X^{[2]}, E_1^{[2]} \otimes \cdots \otimes E_k^{[2]}).$$

Proposition 3.6.16. *Let X be a smooth projective surface. Then the Euler characteristic of tensor products of tautological bundles on $X^{[2]}$ is related to the Euler characteristic of the bundles on X by the formula*

$$\chi_{X^{[2]}}\left(E_1^{[2]} \otimes \cdots \otimes E_k^{[2]}\right)$$

$$= \sum_{\substack{\{P_1, P_2\} \\ \text{partition of } [k]}} \chi\left(\bigotimes_{t \in P_1} E_t\right) \cdot \chi\left(\bigotimes_{t \in P_2} E_t\right) - \sum_{\ell=1}^{k-1} N(k, \ell) \chi\left(S^{\ell-1} \Omega_X \otimes E_1 \otimes \cdots \otimes E_k\right).$$

3.6.4 Long exact sequences on $X_{**}^{[n]}$

Let $k, n \in \mathbb{N}$ and $E_1, \ldots, E_k$ be locally free sheaves on X. For $\hat{M} \subset [k]$, $1 \leq i < j \leq n$, $\ell = 1, \ldots, k$, and $\alpha = 0, \ldots, k - \ell$ with $\ell + \alpha \leq |\hat{M}| =: v$ we set

$$R_\ell^\alpha(\hat{M}; i, j) := \bigoplus_{\substack{M \subset \hat{M}, |M| = \ell + \alpha \\ b: \hat{M} \setminus M \to \{i,j\}}} \wedge^{\ell-1} \varrho_M(b) \quad, \quad \varrho_M(b) := \varrho_M$$

$$R_\ell^{-1}(\hat{M}; i, j) := \bigoplus_{b: \hat{M} \to \{i,j\}} \mathbb{C}(b) \quad, \quad \mathbb{C}(b) = \mathbb{C}.$$

The bijections $u_{[v] \to \hat{M}}$ and $u_{[2] \to \{i,j\}}$ induce for $\alpha = -1, \ldots, v - \ell$ canonical isomorphisms

$$\substack{n \\ k} R_\ell^\alpha(\hat{M}; i, j) \cong \substack{2 \\ v} R_\ell^\alpha$$

with $\substack{2 \\ v} R_\ell^\alpha = {}_v R_\ell^\alpha$ the vector space already defined in subsection 3.6.1. Pulling back the differentials as well as the $\mathfrak{S}_v$-action defined in the last subsection we get $\mathfrak{S}_{\hat{M}}$-equivariant complexes

$$R_\ell^\bullet(\hat{M}; i, j) := \left(0 \to R_\ell^0(\hat{M}; i, j) \to \cdots \to R_\ell^{v-\ell}(\hat{M}; i, j) \to 0\right),$$

$$\tilde{R}_\ell^\bullet(\hat{M}; i, j) := \left(0 \to R_\ell^{-1}(\hat{M}; i, j) \to \cdots \to R_\ell^{v-\ell}(\hat{M}; i, j) \to 0\right).$$

More concretely, the differentials $\tilde{\varphi}(\hat{M};i,j) = d_\ell^{-1}(\hat{M};i,j)\colon R_\ell^{-1}(\hat{M};i,j) \to R_\ell^0(\hat{M};i,j)$ and $d_\ell^\alpha(\hat{M};i,j)\colon R_\ell^\alpha(\hat{M};i,j) \to R_\ell^{\alpha+1}(\hat{M};i,j)$ are given by

$$d_\ell^{-1}(s)(M;b) = \left(\sum_{c\colon M \to \{i,j\}} \varepsilon_c s(b \uplus c) \right) \cdot \zeta_M^1 \wedge \cdots \wedge \zeta_M^{\ell-1},$$

$$d_\ell^\alpha(s)(M;b) = \sum_{m\in M} \varepsilon_{m,M}\big(s(M \setminus \{m\};b, m \mapsto i) - s(M \setminus \{m\};b, m \mapsto j) \big).$$

For every $a\colon [k] \setminus \hat{M} \to [n] \setminus \{i,j\}$ we set

$$H_\ell[\hat{M};i,j;a] := \left(S^{\ell-1}\Omega_X \otimes \bigotimes_{t\in\hat{M}} E_t \right)_{ij} \otimes \bigotimes_{t\in[k]\setminus\hat{M}} \mathrm{pr}_{a(t)}^* E_t$$

and define the complexes $T_\ell^\bullet[\hat{M};i,j;a]$ and $\tilde{T}_\ell^\bullet[\hat{M};i,j;a]$ on X^n by

$$T_\ell^\bullet[\hat{M};i,j;a] := H_\ell[\hat{M};i,j;a] \otimes_{\mathbb{C}} R_\ell^\bullet(\hat{M};i,j),$$
$$\tilde{T}_\ell^\bullet[\hat{M};i,j;a] := H_\ell[\hat{M};i,j;a] \otimes_{\mathbb{C}} \tilde{R}_\ell^\bullet(\hat{M};i,j).$$

Furthermore, we set

$$T_\ell^\bullet := \bigoplus_{\substack{\hat{M}\subset[k],\, 1\leq i<j\leq n \\ a\colon [k]\setminus\hat{M}\to[n]\setminus\{i,j\}}} T_\ell^\bullet[\hat{M};i,j;a], \quad \tilde{T}_\ell^\bullet := \bigoplus_{\substack{\hat{M}\subset[k],\, 1\leq i<j\leq n \\ a\colon [k]\setminus\hat{M}\to[n]\setminus\{i,j\}}} \tilde{T}_\ell^\bullet[\hat{M};i,j;a].$$

For $\alpha \geq 0$ we can also write

$$T_\ell^\alpha = \bigoplus_{\substack{M\subset[k],\, |M|=\ell+\alpha \\ 1\leq i<j\leq n \\ a\colon [k]\setminus\hat{M}\to[n]}} T_\ell^\alpha(M;i,j;a),$$

where $T_\ell^\alpha(M;i,j;a)$ is the direct summand

$$\left(S^{\ell-1}\Omega_X \otimes \bigotimes_{t\in\hat{M}} E_t \right)_{ij} \otimes \bigotimes_{t\in[k]\setminus\hat{M}} \mathrm{pr}_{a(t)}^* E_t \otimes_{\mathbb{C}} \wedge^{\ell-1}\varrho_M(a_{||})$$

of $T_\ell^\alpha(\hat{M}(a);i,j;a_|)$. Here $\hat{M}(a) := M \cup a^{-1}(\{i,j\})$, $a_| := a_{|a^{-1}([n]\setminus\{i,j\})}$, and $a_{||} := a_{|a^{-1}(\{i,j\})}$. We have again $T_\ell \cong T_\ell^0$ given by the isomorphisms $T_\ell(M;i,j;a) \cong T_\ell^0(M;i,j;a)$ induced by

$$\mathbb{C} \xrightarrow{\cong} \wedge^{\ell-1}\varrho_M \quad,\quad 1 \mapsto \zeta_M^1 \wedge \cdots \wedge \zeta_M^{\ell-1}.$$

The $\mathfrak{S}_n$-linearization of T_ℓ^α is given by

$$\lambda_\sigma(s)(M; i, j; a) = \varepsilon_{\sigma,\sigma^{-1}(\{i,j\})}^{\ell+\alpha} \sigma_* s(M; \sigma^{-1}(\{i,j\}); \sigma^{-1} \circ a)$$

where $s(M; i, j; a)$ is the component of s in $T_\ell^\alpha(M; i, j; a)$ and σ_* is the isomorphism induced by flat base change. Furthermore, in the case that all the E_t are equal the $\mathfrak{S}_k$-action is given by

$$(\sigma \cdot s)(M; i, j; a) = \varepsilon_{\sigma,\sigma^{-1}(M)}\sigma \cdot s(\sigma^{-1}(M); i, j; a \circ \sigma)$$

with the σ on the right-hand side being the restriction $\sigma \colon \wedge^{\ell-1} \varrho_{\sigma^{-1}(M)} \to \wedge^{\ell-1}\varrho_M$ of the action of σ on $\wedge^{\ell-1}\varrho_{[k]}$ together with the permutation of the tensor factors. Since the $\mathfrak{S}_n$- and the $\mathfrak{S}_k$-linearization commute they define together a $\mathfrak{S}_n \times \mathfrak{S}_k$-linearization of T_ℓ^α.

Proposition 3.6.17. *For every $\ell = 1, \ldots, k$ the sequences*

$$0 \to K_{\ell**} \to K_{\ell-1**} \xrightarrow{\varphi_{\ell**}} T_{\ell**}^0 \to T_{\ell**}^1 \to \cdots \to T_{\ell**}^{k-\ell} \to 0$$

are exact on X_{**}^n.

Proof. As in the proof of proposition 3.1.5 we will proof the exactness on the open covering given by $U_{ij} = (X_{**}^n \setminus \mathbb{D}) \cup \Delta_{ij}$. Also we may again assume that $E_1 = \cdots = E_k = \mathcal{O}_X$. Since every section $s \in K_0$ over U_{ij} with values $s(a) \in \mathcal{I}_{ij}^{\ell-1}$ for every $a \colon [k] \to [\ell]$ is already a section of $K_{\ell-1} = K_{\ell-1}(i,j)$, we have $\operatorname{im}\varphi_\ell = \operatorname{im}\tilde{\varphi}_\ell$. Thus, it suffices to show that $\tilde{T}_\ell^\bullet$ is cohomologically concentrated in degree -1. There are for $|\hat{M}| = v$ canonical isomorphisms

$$\tilde{T}_\ell^\bullet(\hat{M}; i, j; a) \cong \operatorname{pr}_{ij}^*({}_v^2\tilde{T}_\ell^\bullet) \otimes \bigotimes_{t \in [k]\setminus\hat{M}} \operatorname{pr}_{a(t)}^* E_t.$$

Since ${}_v^2\tilde{T}_\ell^\bullet$ is concentrated in degree -1 by theorem 3.6.6, all the $\tilde{T}_\ell^\bullet(\hat{M}; i, j; a)$ are also. Thus, their direct sum $\tilde{T}_\ell^\bullet$ is indeed concentrated in degree -1. $\qquad\square$

Remark 3.6.18. We can not generalise the above proposition directly to X^n, i.e the sequences

$$0 \to K_\ell \to K_{\ell-1} \xrightarrow{\varphi_\ell} T_\ell^0 \to T_\ell^1 \to \cdots \to T_\ell^{k-\ell} \to 0$$

are not exact. The reason is that for $t \in T_\ell$ a section in the image of φ_ℓ the components $t(M; i, j; a)$ also for various i and j satisfy relations on the intersection of their supports, i.e. on partial diagonals Δ_I with $|I| > 2$. Thus, in a right resolution $S_\ell^\bullet$ of

$$0 \to K_\ell \to K_{\ell-1} \to T_\ell$$

also sheaves supported on Δ_I with $|I| > 2$ will occur (see also subsection 3.7.3). Maybe, there are additional relations on infinitesimal neighbourhoods of partial diagonals coming for example from triples of pairwise diagonals of the form Δ_{ij}, Δ_{jk} and Δ_{ik} (see also subsection 3.7.1). Also, there are relations among the relations. In summary, an exact sequence

$$0 \to K_\ell \to K_{\ell-1} \to T_\ell \to S_\ell^1 \to \cdots \to S_\ell^N \to 0 \,.$$

will most likely look very complicated for general k and n.

3.7 Tensor products of tautological bundles on $X^{[n]}$

In this section we will enlarge the exact sequences

$$0 \to K_\ell^{\mathfrak{S}_n} \to K_{\ell-1}^{\mathfrak{S}_n} \xrightarrow{\varphi_\ell^{\mathfrak{S}_n}} T_\ell^{\mathfrak{S}_k}$$

for $\ell = 1, \ldots, k-1$ to long exact sequences on $S^n X$ with a zero on the right in the cases $k = 2, 3$. This will lead to results for the cohomology of double and triple tensor products of tautological bundles. The results for the double tensor product can be found in [Sca09b] for a bigger class of pairs of sheaves E_1, E_2 on X.

3.7.1 Restriction of local sections to closed subvarieties

Lemma 3.7.1. *Let X be a quasi-projective variety. Then X^n has an open covering consisting of subsets of the form U^n for $U \subset X$ open and affine.*

Proof. See [Sca09a, Lemma 1.4.3]. □

Lemma 3.7.2. *Let X be a scheme, $Z_1, Z_2 \subset X$ two closed subschemes and $Z_1 \cap Z_2$ their scheme-theoretic intersection. Then for every locally free sheaf F on X the following sequence is exact*

$$F \to F_{|Z_1} \oplus F_{|Z_2} \to F_{|Z_1 \cap Z_2} \to 0$$
$$s \mapsto \begin{pmatrix} s_{|Z_1} \\ s_{|Z_2} \end{pmatrix}, \ \begin{pmatrix} a \\ b \end{pmatrix} \mapsto a_{|Z_1 \cap Z_2} - b_{|Z_2 \cap Z_2} \,.$$

Proof. Since the question is local, we can assume that F is a trivial vector bundle. Since in this case the restriction of sections is defined component-wise, we can assume that $F = \mathcal{O}_X$ is the trivial line bundle. Furthermore we can assume that $X = \operatorname{Spec} A$ is affine. Now the assertion follows from the fact that for two ideals $I, J \subset A$ sequence

$$A \to A/I \oplus A/J \to A/(I+J) \to 0$$

is exact. □

Let $I \subset [n]$ and $i \in [n] \setminus I$. Then the closed embedding $\iota \colon \Delta_{I \cup \{i\}} \hookrightarrow \Delta_I$ induces by the universal property of the $\mathfrak{S}_{\overline{I \cup \{i\}}}$-quotient $\Delta_{I \cup \{i\}} \times S^{\overline{I \cup \{i\}}} X$ the commutative diagram

$$
\begin{array}{ccc}
\Delta_{I \cup \{i\}} & \xrightarrow{\ \iota\ } & \Delta_I \\
{\scriptstyle \pi_{\overline{I \cup \{i\}}}} \big\downarrow & & \big\downarrow {\scriptstyle \pi_{\overline{I}}} \\
\Delta_{I \cup \{i\}} \times S^{\overline{I \cup \{i\}}} X & \xrightarrow{\ \bar{\iota}\ } & \Delta_I \times S^{\overline{I}} X.
\end{array}
$$

We also allow the case that $I = \{j\}$ consists of only one element. Then $\Delta_I = X^n$ and $\Delta_I \times S^{\overline{I}} X = X^I \times S^{\overline{I}} X \cong X \times S^{n-1} X$.

Lemma 3.7.3. *The morphism $\bar{\iota}$ is again a closed embedding.*

Proof. Let $|I| = \ell$. We can make the identifications $\Delta_{I \cup \{i\}} \times S^{\overline{I \cup \{i\}}} X \cong X \times S^{n-\ell-1}$ and $\Delta_I \times S^{\overline{I \cup \{i\}}} X \cong X \times S^{n-\ell}$. Under these identifications the map $\bar{\iota}$ on the level of points is given by $(x, \Sigma) \mapsto (x, x + \Sigma)$. Hence, it is injective. Since the quotient morphism $\pi_{\overline{I}}$ is finite and hence proper, $\mathrm{im}(\bar{\iota}) = \pi_{\overline{I}}(\Delta_{I \cup \{i\}})$ is a closed subset. It remains to show that the morphism of sheaves $\mathcal{O}^{\mathfrak{S}_{\overline{I}}}_{\Delta_I} \to \mathcal{O}^{\mathfrak{S}_{\overline{I \cup \{i\}}}}_{\Delta_{U \cup \{i\}}}$ is still surjective. This can be seen by covering X^n by open affines of the form U^n (see lemma 3.7.1) and using the lemmas 5.1.1. and 5.1.2 of [Sca09a] in the case $F = \mathcal{O}_X$ and $* = 0$. Another way to prove it is the following. We assume for simplicity $I = [\ell]$ and $i = \ell + 1$. For $\ell < k \leq n$ we set $Z_k := \cup_{j=\ell+1}^{k} \Delta_{[\ell] \cup \{j\}} \subset \Delta_{[\ell]}$ and $Z := Z_n$. Now for a $\mathfrak{S}_{\overline{[\ell+1]}}$-invariant local regular function $s \in \mathcal{O}_{\Delta_{[\ell+1]}}$ we define $\tilde{s} \in \mathcal{O}_Z$ by $\tilde{s}_{|\Delta_{[\ell+1]}} = s$ and $\tilde{s}_{|\Delta_{[\ell] \cup \{j\}}} = (\ell + 1, j)_* s$ for $j \in [\ell+2, n]$. Since every permutation $\sigma \in \mathfrak{S}_{\overline{[\ell]}}$ can be written in the form $\sigma = (\ell + 1, j) \circ \mu$ with $j \in [\ell + 1, n]$ and $\mu \in \mathfrak{S}_{\overline{[\ell+1]}}$, the function $\tilde{s}$ is $\mathfrak{S}_{\overline{[\ell]}}$-invariant. Thus, there is a $t \in \mathcal{O}^{\mathfrak{S}_{[\ell]}}_{\Delta_{[\ell]}}$ whose restriction to Z is $\tilde{s}$. In particular $t_{|\Delta_{[\ell+1]}} = s$. What is left to show is that $\tilde{s}$ is indeed a regular function on Z using lemma 3.7.2. We show by induction over k that it is a regular function on every Z_k. The case $k = \ell + 1$ is clear. Hence, we assume that $\tilde{s}_{|Z_{k-1}}$ is regular and $k \geq \ell + 2$. We show that the equation

$$
Z_{k-1} \cap \Delta_{[\ell] \cup \{k\}} = (\cup_{j=\ell+1}^{k-1} \Delta_{[\ell] \cup \{j\}}) \cap \Delta_{[\ell] \cup \{k\}} = \cup_{j=\ell+1}^{k-1} (\Delta_{[\ell] \cup \{j\}} \cap \Delta_{[\ell] \cup \{k\}})
$$

holds scheme-theoretically which means that there is the equation of ideals

$$
(\cap_{j=\ell+1}^{k-1} \mathcal{I}_{[\ell] \cup \{j\}}) + \mathcal{I}_{[\ell] \cup \{k\}} = \cap_{j=\ell+1}^{k-1} (\mathcal{I}_{[\ell] \cup \{j\}} + \mathcal{I}_{[\ell] \cup \{k\}}).
$$

We can test the equality stalk-wise and thus take local coordinates as in the proof of 1.5.20 such that

$$
\mathcal{I}_{[\ell] \cup \{j\}} = \big(x_1 - x_2, \ldots, x_{\ell-1} - x_\ell, x_\ell - x_j, y_1 - y_2, \ldots, y_{\ell-1} - y_\ell, y_\ell - y_j\big).
$$

Then one can compute that both sides of the above equation equal the ideal

$$\left(x_1 - x_2, \ldots, x_{\ell-1} - x_\ell, \prod_{j=\ell+1}^{k-1} (x_\ell - x_j), x_\ell - x_k, y_1 - y_2, \ldots, y_{\ell-1} - y_\ell, \prod_{j=\ell+1}^{k-1} (y_\ell - y_j), y_\ell - y_k\right).$$

Now for every $j = \ell + 1, \ldots, k - 1$ the functions $(\ell + 1, j)_*s$ and $(\ell + 1, k)_*s$ coincide on $\Delta_{[\ell] \cup \{j\}} \cap \Delta_{[\ell] \cup \{k\}} = \Delta_{[\ell] \cup \{j,k\}}$. Thus, $(\ell + 1, k)_*s$ and $\tilde{s}_{Z_{k-1}}$ coincide on $Z_{k-1} \cap \Delta_{[\ell] \cup \{k\}}$ and we can indeed apply lemma 3.7.2 to get that $\tilde{s}$ is a regular function on Z_k. $\square$

Definition 3.7.4. Let a finite group G act on a scheme X. A locally free G-sheaf F on X of rank r is called *locally trivial* (as a G-sheaf) if there is a cover over X consisting of G-invariant open subsets $U \subset X$ such that $F|_U \cong_G \mathcal{O}_U^{\oplus r}$. The G-linearization of $\mathcal{O}_U$ is the natural one (see subsection 1.4.9).

Remark 3.7.5. Let F be a locally free sheaf on X and $\emptyset \neq I \subset [n]$. Then F_I (see subsection 1.2.4) is locally trivial as a $\overline{\mathfrak{S}_I}$-sheaf on Δ_I.

Corollary 3.7.6. *Let $\mathcal{F}$ be a locally trivial $\mathfrak{S}_{\bar{I}}$-sheaf on Δ_I and $r: \mathcal{F} \to \mathcal{F}_{|\Delta_{I \cup \{i\}}}$ the morphism given by restriction of sections. Then the induced morphism $\underline{r}: \mathcal{F}^{\mathfrak{S}_I} \to \mathcal{F}_{|\Delta_{I \cup \{i\}}}^{\overline{\mathfrak{S}_{I \cup \{i\}}}}$ is still surjective.*

Proof. The morphism is given by restricting sections of the locally free sheaf $\mathcal{F}^{\mathfrak{S}_I}$ along the closed embedding $\bar{\iota}$ (see also remark 3.2.5). $\square$

3.7.2 Double tensor products

We consider the case $k = 2$ and $n \geq 2$. Then by corollary 3.2.4 we have

$$\mu_*(E_1^{[n]} \otimes E_2^{[n]}) \cong K_1^{\mathfrak{S}_n} = \ker(\varphi_1^{\mathfrak{S}_n}).$$

We have $J_0(1) = \{(1,2), (1,1)\}$ and $\hat{J}_1(1) = \{(\{1\}; 1), (\{2\}; 1)\}$ (see remark 3.2.1). Both $T_1(\{1\}; 1)$ and $T_1(\{2\}; 1)$ equal $(E_1 \otimes E_2)_{12}$. Let $\tau = (1\ 2) \in \mathfrak{S}_n$. For every $s \in K_0^{\mathfrak{S}_n}$ we have

$$\varphi_1(s)(\{1\}; 1) = s(1,1)_{|\Delta_{12}} - s(2,1)_{|\Delta_{12}} = s(1,1)_{|\Delta_{12}} - \tau_* s(1,2)_{|\Delta_{12}}$$
$$= s(1,1)_{|\Delta_{12}} - s(1,2)_{|\Delta_{12}}$$
$$= \varphi_1(s)(\{2\}; 1).$$

Thus $\ker(\varphi_1^{\mathfrak{S}_n}) = \ker(\varphi_1^{\mathfrak{S}_n}(\{2\}; 1))$. The morphism $\varphi_1(\{2\}; 1): K_0 \to T_1(\{2\}; 1, 2; 1)$ is given by restricting sections to the closed subscheme Δ_{12}. Hence, by corollary 3.7.6 the morphism $\varphi_1^{\mathfrak{S}_n}(\{2\}; 1)$ is still surjective. With these considerations we get the following theorem which was already proven by Scala in [Sca09a] and [Sca09b].

Theorem 3.7.7. *On $S^n X$ there is the exact sequence*

$$0 \to \mu_*(E_1^{[n]} \otimes E_2^{[n]}) \to \begin{array}{c} \pi_*(p_1^* E_1 \otimes p_2^* E_2)^{\mathfrak{S}_{[3,n]}} \\ \oplus \\ \pi_* p_1^*(E_1 \otimes E_2)^{\mathfrak{S}_{[2,n]}} \end{array} \xrightarrow{\varphi_1^{\mathfrak{S}_n}(\{2\};1)} \pi_*(E_1 \otimes E_2)_{12}^{\mathfrak{S}_{[3,n]}} \to 0 \,.$$

Corollary 3.7.8. *Let X be projective. Then*

$$\chi\left(E_1^{[n]} \otimes E_2^{[n]}\right) = \chi(E_1) \cdot \chi(E_2) \cdot \binom{\chi(\mathcal{O}_X) + n - 3}{n-2} + \chi(E_1 \otimes E_2) \cdot \binom{\chi(\mathcal{O}_X) + n - 3}{n-1}$$

Proof. By lemma 3.2.7 and theorem 3.7.7 it follows indeed that

$$
\begin{aligned}
\chi(E_1^{[n]} \otimes E_2^{[n]}) =& \chi(E_1) \cdot \chi(E_2) \cdot \binom{\chi(\mathcal{O}_X) + n - 3}{n-2} \\
&+ \chi(E_1 \otimes E_2) \cdot \left(\binom{\chi(\mathcal{O}_X) + n - 2}{n-1} - \binom{\chi(\mathcal{O}_X) + n - 3}{n-2} \right) \\
=& \chi(E_1) \cdot \chi(E_2) \cdot \binom{\chi(\mathcal{O}_X) + n - 3}{n-2} + \chi(E_1 \otimes E_2) \cdot \binom{\chi(\mathcal{O}_X) + n - 3}{n-1} \,.
\end{aligned}
$$

$\square$

Theorem 3.7.9. *For every $i \geq 0$ there is a natural isomorphism*

$$\mathrm{H}^i(X^{[n]}, E_1^{[n]} \otimes E_2^{[n]}) \cong \ker\left(\mathrm{H}^i(\varphi_1(\{2\}; 1))^{\mathfrak{S}_n} \right) \,.$$

Proof. In section 5.1 of [Sca09a] it is shown that the component

$$\mathrm{H}^i(\varphi_1^{\mathfrak{S}_n}((1,1),(\{2\};1))\colon \left(\mathrm{H}^*(E_1 \otimes E_2) \otimes S^{n-1}\,\mathrm{H}^*(\mathcal{O}_X) \right)^i \to \left(\mathrm{H}^*(E_1 \otimes E_2) \otimes S^{n-2}\,\mathrm{H}^*(\mathcal{O}_X) \right)^i$$

is surjective for every $i \in \mathbb{N}$. Hence, the long exact cohomology sequence associated to the short exact sequence of theorem 3.7.7 splits into short exact sequences for every $i \in \mathbb{N}$ which yields the result. $\square$

3.7.3 Triple tensor products

We consider the case $k = 3$ and $n \geq 3$. We have by corollary 3.2.4

$$\mu_*(E_1^{[n]} \otimes E_2^{[n]} \otimes E_3^{[n]}) \cong K_2^{\mathfrak{S}_n} \,.$$

We will successively describe $K_1^{\mathfrak{S}_n}$ and $K_2^{\mathfrak{S}_n}$ as kernels of exact sequences with 0 on the right. We consider the set of representatives

$$\tilde{I} := \big\{(1,1,1),\,(1,1,3),\,(1,3,1),\,(3,1,1),\,(1,2,3)\big\}$$

of the $\mathfrak{S}_n$-orbits of I_0 and the set of representatives $\tilde{J} = \tilde{J}^1 \cup \tilde{J}^2 \cup \tilde{J}^3 \cup \tilde{J}^4$ of the $\mathfrak{S}_n$-orbits of $\hat{I}_1$ given by

$$\tilde{J}^1 = \{(\{1\};1,3;(1,1)),\,(\{2\};1,3;(1,1)),\,(\{3\};1,3;(1,1))\},$$
$$\tilde{J}^2 = \{(\{1\};1,3;(1,3)),\,(\{2\};1,3;(3,1)),\,(\{3\};1,3;(1,3))\},$$
$$\tilde{J}^3 = \{(\{1\};1,2;(1,3)),\,(\{2\};1,2;(3,1)),\,(\{3\};1,2;(1,3))\},$$
$$\tilde{J}^4 = \{(\{1\};1,2;(3,1)),\,(\{2\};1,2;(1,3)),\,(\{3\};1,2;(3,1))\}.$$

Here in the tuple $(\{t\};i,j;(\alpha,\beta))$ the canonical identification $\mathrm{Map}([3]\setminus\{t\},[n]) \cong [n]^2$ is used, i.e. the tuple (α,β) stands for the map $a\colon [3]\setminus\{t\} \to [n]$ given by $a(\min([3]\setminus\{t\})) = \alpha$ and $a(\max([3]\setminus\{t\})) = \beta$. One can check that $\tilde{I}$ and $\tilde{J}$ are indeed systems of representatives of the $\mathfrak{S}_n$-orbits of I_0 respectively $\hat{I}_1$ by giving bijections which preserve the orbits $\tilde{I} \cong J_0(1)$ and $\tilde{J} \cong \hat{J}_1(1)$ to the sets of representatives given in remark 3.2.1. Thus, we have the isomorphism $K_0^{\mathfrak{S}_n} \cong \oplus_{a\in\tilde{I}} K_0(a)^{\mathfrak{S}_{\overline{\mathrm{im}(a)}}}$ and $T_1^{\mathfrak{S}_n}$ is given by

$$\bigoplus_{\tilde{J}^1} T_\ell(\{t\};1,3;(\alpha,\beta))^{\mathfrak{S}_{\overline{\{1,3\}}}} \oplus \bigoplus_{\tilde{J}^2} T_\ell(\{t\};1,3;(\alpha,\beta))^{\mathfrak{S}_{\overline{\{1,3\}}}} \oplus \bigoplus_{\tilde{J}^3\cup\tilde{J}^4} T_\ell(\{t\};1,2;(\alpha,\beta))^{\mathfrak{S}_{\overline{[4,n]}}}.$$

We use for $r = 1,2,3,4$ the notation

$$T_\ell(r) = \bigoplus_{(\{t\};i,j;(\alpha,\beta))\in\tilde{J}^r} T_1(\{t\};i,j;(\alpha,\beta))$$

as well as $\varphi_1(r) = \oplus_{\tilde{J}^r} \varphi_1(\{t\};i,j;(\alpha,\beta))$. Then

$$K_1^{\mathfrak{S}_n} = \ker(\varphi_1^{\mathfrak{S}_n}) = \ker(\varphi_1^{\mathfrak{S}_n}(1)) \cap \ker(\varphi_1^{\mathfrak{S}_n}(2)) \cap \ker(\varphi_1^{\mathfrak{S}_n}(3)) \cap \ker(\varphi_1^{\mathfrak{S}_n}(4)).$$

For every $\gamma \in \tilde{J}_1 \cup \tilde{J}_2$ the sheaves $T_1(\gamma)$ are canonically isomorphic to

$$\mathcal{F} := \big(E_1 \otimes E_2 \otimes E_3\big)_{13}.$$

Hence, $T_1(2)^{\mathfrak{S}_{\overline{\{1,3\}}}} \cong T_1(1)^{\mathfrak{S}_{\overline{\{1,3\}}}} \cong (\mathcal{F}^{\mathfrak{S}_{\overline{\{1,3\}}}})^{\oplus 3}$. Moreover, for every $a = (1,1,3),(1,3,1),(3,1,1)$ the restriction of the sheaf $K_0(a)$ to Δ_{13} is canonically isomorphic to $\mathcal{F}$.

Lemma 3.7.10. *The following sequence on $S^n X$ is exact:*

$$0 \to \ker(\varphi_1^{\mathfrak{S}_n}(1)) \cap \ker(\varphi_1^{\mathfrak{S}_n}(2)) \to K_0^{\mathfrak{S}_n} \xrightarrow{\varphi_1^{\mathfrak{S}_n}(1)} T_1(1)^{\mathfrak{S}_{\overline{\{1,3\}}}} \to 0 \,.$$

In particular $\ker(\varphi_1^{\mathfrak{S}_n}(1)) \cap \ker(\varphi_1^{\mathfrak{S}_n}(2)) = \ker(\varphi_1^{\mathfrak{S}_n}(1))$.

Proof. Let $\tau = (1\ 3) \in \mathfrak{S}_n$. Note that for every $s \in K_0^{\mathfrak{S}_n}$ a $\mathfrak{S}_n$-invariant section $s(3,3,1) = \tau_* s(1,1,3)$ holds. Since τ acts trivially on Δ_{13}, there is the equality

$$s(3,3,1)_{|\Delta_{13}} = s(1,3,3)_{|\Delta_{13}}$$

in $\mathcal{F}$. Analogously, $s(3,1,3)_{|\Delta_{13}} = s(1,3,1)_{|\Delta_{13}}$ and $s(1,1,3)_{|\Delta_{13}} = s(3,3,1)_{|\Delta_{13}}$. The exactness in the first two degrees comes from the inclusion $\ker(\varphi_1(1)) \subset \ker(\varphi_1(2))$. Indeed, a section $s \in K_0^{\mathfrak{S}_n}$ is in $\ker(\varphi_1^{\mathfrak{S}_n}(1))$ if and only if

$$s(1,1,1)_{|\Delta_{13}} = s(1,1,3)_{|\Delta_{13}} = s(1,3,1)_{|\Delta_{13}} = s(3,1,1)_{|\Delta_{13}} \,,$$

whereas $s \in \ker(\varphi_1(2))$ holds if and only if

$$s(1,1,3)_{|\Delta_{13}} = s(1,3,1)_{|\Delta_{13}} = s(3,1,1)_{|\Delta_{13}} \,.$$

For the exactness on the right consider $t \in T_1(1)^{\mathfrak{S}_{\overline{\{1,3\}}}}$. Then locally there are sections $s(a) \in K_0(a)^{\mathfrak{S}_{\overline{\{1,3\}}}}$ for $a = (1,1,3),(1,3,1),(3,1,1)$ such that

$$s(3,1,1)_{\Delta_{13}} = t(\{1\};1,3;(1,1)) \quad , \quad s(1,3,1)_{\Delta_{13}} = t(\{2\};1,3;(1,1)),$$
$$s(1,1,3)_{\Delta_{13}} = t(\{3\};1,3;(1,1)) \,.$$

By setting $s(1,1,1) = 0 = s(1,2,3)$ we have indeed defined a local section $s \in K_0^{\mathfrak{S}_n}$ with $\varphi_1^{\mathfrak{S}_n}(1)(s) = t$. $\qquad\square$

For every tuple $(\{t\};i,j;(\alpha,\beta)) \in \tilde{J}$ the restriction of the corresponding sheaf to Δ_{123} is given by

$$T_1(\{t\};i,j;(\alpha,\beta))_{|\Delta_{123}} = (E_1 \otimes E_2 \otimes E_3)_{123} =: \mathcal{E} \,.$$

Thus we can define the $\mathfrak{S}_{[n,4]}$-equivariant morphism

$$F: T_1(3) \to \mathcal{E}^{\oplus 2} \,, \quad s \mapsto \begin{pmatrix} s(\{1\};1,2;(1,3))_{|\Delta_{123}} - s(\{2\};1,2;(3,1))_{|\Delta_{123}} \\ s(\{2\};1,2;(3,1))_{|\Delta_{123}} - s(\{3\};1,2;(1,3))_{|\Delta_{123}} \end{pmatrix} \,.$$

Proposition 3.7.11. *On $S^n X$ there is the exact sequence*

$$0 \to K_1^{\mathfrak{S}_n} \to \ker(\varphi_1^{\mathfrak{S}_n}(1)) \xrightarrow{\varphi_1^{\mathfrak{S}_n}(3)} T_1(3)^{\mathfrak{S}_{[4,n]}} \xrightarrow{F^{\mathfrak{S}_{[4,n]}}} (\mathcal{E}^{\oplus 2})^{\mathfrak{S}_{[4,n]}} \to 0 \,.$$

Proof. Because of the invariance under the transposition $(1\ 2)$, we have for every $\in K_0^{\mathfrak{S}_n}$ the equality $s(1,2,3)_{|\Delta_{12}} = s(2,1,3)_{|\Delta_{12}}$. Thus,

$$\varphi_1(s)(\{1\};1,2;(1,3)) = s(1,1,3)_{|\Delta_{12}} - s(2,1,3)_{|\Delta_{12}} = s(1,1,3)_{|\Delta_{12}} - s(1,2,3)_{|\Delta_{12}}$$
$$= \varphi_1(s)(\{2\};1,2;(1,3))\,.$$

Similarly, we get

$$\varphi_1(s)(\{2\};1,2;(3,1)) = \varphi_1(s)(\{3\};1,2;(3,1))\,,\ \varphi_1(s)(\{3\};1,2;(1,3)) = \varphi_1(s)(\{1\};1,2;(3,1))\,.$$

This shows that $\ker(\varphi_1^{\mathfrak{S}_n}(3)) = \ker(\varphi_1^{\mathfrak{S}_n}(4))$. Together with lemma 3.7.10 we thus have $K_1^{\mathfrak{S}_n} = \ker(\varphi_1^{\mathfrak{S}_n}(1)) \cap \ker(\varphi_1^{\mathfrak{S}_n}(3))$ which shows the exactness in the first two degrees. Let $s \in K_0$ be a $\mathfrak{S}_n$-invariant section with $\varphi_1(1)(s) = 0$, i.e.

$$s(1,1,1)_{\Delta_{13}} = s(1,1,3)_{\Delta_{13}} = s(1,3,1)_{\Delta_{13}} = s(3,1,1)_{\Delta_{13}}\,.$$

By the $\mathfrak{S}_n$-invariance we have

$$s(2,1,3) = (1\ 2)_* s(1,2,3)\,,\ s(3,2,1) = (1\ 3)_* s(1,2,3)\,,\ s(1,3,2) = (2\ 3)_* s(1,2,3)\,.$$

Since $\mathfrak{S}_{[3]}$ acts trivially on Δ_{123}, it follows that

$$s(1,2,3)_{|\Delta_{123}} = s(2,1,3)_{|\Delta_{123}} = s(3,2,1)_{|\Delta_{123}} = s(1,3,2)_{|\Delta_{123}}\,.$$

This yields for the first component F_1 of the morphism F

$$F_1(\varphi_1(3)(s)) = \varphi_1(s)(\{1\};1,2;(1,3))_{\Delta_{123}} - \varphi_1(s)(\{2\};1,2;(3,1))_{\Delta_{123}}$$
$$= \big(s(1,1,3) - s(2,1,3) - s(3,1,1) + s(3,2,1)\big)_{|\Delta_{123}} = 0\,.$$

Analogously, $F_2(\varphi_1(3)(s)) = 0$ which gives the inclusion $\mathrm{im}(\varphi_1^{\mathfrak{S}_n}(3)) \subset \ker(F^{\mathfrak{S}_{[4,n]}})$. To show the other inclusion let $t \in \ker(F^{\mathfrak{S}_{[4,n]}})$, i.e.

$$t(\{1\};1,2;(1,3))_{|\Delta_{123}} = t(\{2\};1,2;(3,1))_{|\Delta_{123}} = t(\{3\};1,2;(1,3))_{|\Delta_{123}} =: t_{123}\,.$$

As explained in remark 3.2.5 we consider the invariants of the occurring direct summands by their stabilisers as sheaves on the quotients of X^n by the stabilisers, i.e. we have for $a = (1,1,3),(1,3,1),(3,1,1)$, $\gamma_1 \in \tilde{J}^1$, and $\gamma_3 \in \tilde{J}^3$

$$\mathcal{E}^{\mathfrak{S}_{[4,n]}} \in \mathrm{Coh}(\Delta_{123} \times S^{[4,n]}X)\,,\ T_1(\gamma_3)^{\mathfrak{S}_{[4,n]}} \in \mathrm{Coh}(\Delta_{12} \times S^{[4,n]}X)$$
$$\mathcal{F}^{\mathfrak{S}_{\overline{\{1,3\}}}} = T_1(\gamma_1)^{\mathfrak{S}_{\overline{\{1,3\}}}} \in \mathrm{Coh}(\Delta_{13} \times S^{\overline{\{1,3\}}}X)\,,\ K_0(a)^{\mathfrak{S}_{\overline{\{1,3\}}}} \in \mathrm{Coh}(X^{\{1,3\}} \times S^{\overline{\{1,3\}}}X)$$

There are the two cartesian diagrams of closed embeddings (see lemma 3.7.3)

$$
\begin{array}{ccc}
\Delta_{123} & \longrightarrow & \Delta_{12} \\
\downarrow & & \downarrow \\
\Delta_{13} & \longrightarrow & X^n
\end{array}
\quad , \quad
\begin{array}{ccc}
\Delta_{123} \times S^{[4,n]}X & \longrightarrow & \Delta_{12} \times S^{[4,n]}X \\
\downarrow & & \downarrow \\
\Delta_{13} \times S^{\overline{\{1,3\}}}X & \longrightarrow & X^{\{1,3\}} \times S^{\overline{\{1,3\}}}X
\end{array}
$$

where the second one is induced by the first one by the universal properties of the quotient morphisms. That the second diagram is indeed again cartesian can be seen by assuming that X is affine with $\mathcal{O}(X) = A$ (see lemma 3.7.1). Then

$$
\begin{aligned}
& \mathcal{O}(\Delta_{12} \times S^{[4,n]}X) \otimes_{\mathcal{O}(X^{\{1,3\}} \times S^{\overline{\{1,3\}}}X)} \mathcal{O}(\Delta_{13} \times S^{\overline{\{1,3\}}}X) \\
=& (A^{\otimes 2} \otimes S^{n-3}A) \otimes_{A^{\otimes 2} \otimes S^{n-2}A} (A \otimes S^{n-2}A) \cong (A^{\otimes 2} \otimes S^{n-3}A) \otimes_{A^{\otimes 2}} A \cong A \otimes S^{n-3}A \\
=& \mathcal{O}(\Delta_{123} \times S^{[4,n]}X) .
\end{aligned}
$$

We now choose $s_{13} \in \mathcal{F}^{\mathfrak{S}_{\overline{\{1,3\}}}}$ as any local section on $\Delta_{13} \times S^{\overline{\{1,3\}}}X$ such that

$$
s_{13|\Delta_{123} \times S^{[4,n]}X} = t_{123} .
$$

Then by lemma 3.7.2 there exists a local section $s(1,1,3) \in K_0(1,1,3)^{\mathfrak{S}_{\overline{\{1,3\}}}}$ on the quotient $X^{\{1,3\}} \times S^{\overline{\{1,3\}}}X$ such that

$$
s(1,1,3)_{|\Delta_{13} \times S^{\overline{\{1,3\}}}X} = s_{13} \quad , \quad s(1,1,3)_{|\Delta_{12} \times S^{[4,n]}X} = t(\{1\};1,2;(1,3)) .
$$

Analogously, there are local sections $s(1,3,1) \in K_0(1,3,1)^{\mathfrak{S}_{\overline{\{1,3\}}}}$ and $s(3,1,1) \in K_0(3,1,1)^{\mathfrak{S}_{\overline{\{1,3\}}}}$ such that their restrictions to the appropriate closed subvarieties are s_{13} and $t(\{3\};1,2;(1,3))$ respectively $t(\{2\};1,2;(3,1))$. We furthermore set $s(1,2,3) = 0$ and choose any section $s(1,1,1) \in K_0(1,1,1)$ on $X^1 \times S^{[2,n]}$ that restricts to s_{13} on $\Delta_{13} \times S^{\overline{\{1,3\}}}X$. Then $s = (s(a))_{a \in \bar{I}}$ is indeed a section of $\ker(\varphi_1^{\mathfrak{S}_n}(1))$ with the property that $\varphi_1(3)(s) = t$. Finally, the surjectivity of the morphism F follows directly from its definition. This implies the exactness on the right. $\qquad\square$

A system of representatives of $\hat{I}_2$ is given by

$$
\tilde{J}_2 := \{(\{1,2\};1,3;1), (\{1,3\};1,3;1), (\{2,3\};1,3;1)\}
$$

The sheaves $T_2(\gamma)$ for $\gamma \in \tilde{J}_2$ are all canonically isomorphic to

$$
\mathcal{H} := N_{\Delta_{13}} \otimes (E_1 \otimes E_2 \otimes E_3)_{13} \cong (\Omega_X \otimes E_1 \otimes E_2 \otimes E_3)_{13} .
$$

The restriction $\mathcal{H} \to \mathcal{H}_{|\Delta_{123}}$ induces the morphism res: $\mathcal{H}^{\mathfrak{S}_{\overline{\{1,3\}}}} \to \mathcal{H}^{\mathfrak{S}_{[4,n]}}_{|\Delta_{123}}$.

Proposition 3.7.12. *The following sequence is exact:*

$$0 \to K_2^{\mathfrak{S}_n} \to K_1^{\mathfrak{S}_n} \xrightarrow{\varphi_2^{\mathfrak{S}_n}(\{1,2\};1,3;1)} \mathcal{H}^{\mathfrak{S}_{\overline{\{1,3\}}}} \xrightarrow{\text{res}} \mathcal{H}^{\mathfrak{S}_{[4,n]}}_{|\Delta_{123}} \to 0.$$

Proof. Let $s \in K_1^{\mathfrak{S}_n}$ and $\tau = (1\ 3) \in \mathfrak{S}_n$. Then

$$s(3,3,1) = \tau_* s(1,1,3), \ s(3,1,3) = \tau_* s(1,3,1), \ s(1,3,3) = \tau_* s(3,1,1).$$

Since τ acts by $(-1)^{2+1} = -1$ (see remark 3.1.2) on $\mathcal{H}$ one can compute that $\varphi_2(s)(\gamma)$ is equal for all $\gamma \in \tilde{J}_2$. Namely, we have modulo $\mathcal{I}_{13}^2$ the equalities

$$
\begin{aligned}
\varphi_2(s)(\{1,2\};1,3;1) &= s(1,1,1) - s(3,1,1) - s(1,3,1) + \tau_* s(1,1,3) \\
&= s(1,1,1) - s(3,1,1) - \tau_*\big(-s(1,1,3) + \tau_* s(1,3,1)\big) \\
&= s(1,1,1) - s(3,1,1) - s(1,3,1) + \tau_* s(1,3,1) \\
&= \varphi_2(s)(\{1,3\};1,3;1) = \cdots = \varphi_2(s)(\{2,3\};1,3;1).
\end{aligned}
$$

Thus, $K_2^{\mathfrak{S}_n} = \ker(\varphi_2^{\mathfrak{S}_n}) = \ker(\varphi_2^{\mathfrak{S}_n}(\gamma))$ for any $\gamma \in \tilde{J}_2$. In particular, this holds for the tuple $\gamma = (\{1,2\};1,3;1)$ which shows the exactness of the sequence in the first two degrees. The surjectivity of the map res is due to the fact that it is given by the restriction of sections along the closed embedding $\Delta_{123} \times S^{[4,n]} X \to \Delta_{13} \times S^{\overline{\{1,3\}}} X$ which is induced by the embedding $\Delta_{123} \hookrightarrow \Delta_{13}$. Thus, it is only left to show that the sequence is exact at the term $\mathcal{H}^{\mathfrak{S}_{\overline{\{1,3\}}}}$ which equivalent to the equality

$$\operatorname{im}(\varphi_2^{\mathfrak{S}_n}(\{1,2\};1,3;1)) = (I_{123} \cdot \mathcal{H})^{\mathfrak{S}_{\overline{\{1,3\}}}}.$$

Since it suffices to show the equality on a family of open subsets of X^n covering Δ_{123}, we can assume that $E_1 = E_2 = E_3 = \mathcal{O}_X$ (see proof of proposition 3.1.5), i.e. $\mathcal{H} = \mathcal{I}_{13}/\mathcal{I}_{13}^2 = N_{\Delta_{13}}$. Furthermore, we may assume that $\mathcal{I}_{13}/\mathcal{I}_{13}^2$ is a free $\mathcal{O}_{\Delta_{13}}$-module with generators $\bar{\zeta}_1, \bar{\zeta}_2$. The generators $\bar{\zeta}_i$ can be taken as the pullback along pr_{13} of generators of N_Δ where Δ is the diagonal in X^2. Thus, we may assume that their representatives $\zeta_i \in \mathcal{I}_{13}$ are $\mathfrak{S}_{\overline{\{1,3\}}}$-invariant. For $f \in \mathcal{I}_{123}^{\mathfrak{S}_{\overline{\{1,3\}}}}$ and $i = 1,2$ we have to show that $f \cdot \bar{\zeta}_i \in \operatorname{im}(\varphi_2^{\mathfrak{S}_n}(\{1,2\};1,3;1))$. We set $F = -f \cdot \zeta_i \in \mathcal{I}_{123} \cdot \mathcal{I}_{13}$. Then also $(1\ 3)_* F \in \mathcal{I}_{123} \cdot \mathcal{I}_{13}$. Since

$$
\begin{aligned}
\mathcal{I}_{123} \mathcal{I}_{13} = (\mathcal{I}_{12} + \mathcal{I}_{23})\mathcal{I}_{13} = \mathcal{I}_{12}\mathcal{I}_{13} + \mathcal{I}_{23}\mathcal{I}_{13} &\subset \mathcal{I}(\Delta_{12} \cup \Delta_{13}) + \mathcal{I}_{23} \\
&= \mathcal{I}((\Delta_{12} \cup \Delta_{13}) \cap \Delta_{23}),
\end{aligned}
$$

the restriction $((1\ 3)_* F)_{|(\Delta_{12} \cup \Delta_{13}) \cap \Delta_{23}}$ vanishes. Here $\cup$ and $\cap$ denote the scheme-theoretic union and intersection. By lemma 3.7.2 there is a $G \in \mathcal{O}_{X^n}$ such that $G_{|\Delta_{23}} = (1\ 3)_* F$ and

$G_{|\Delta_{12}} = 0 = G_{|\Delta_{13}}$. Since 0 as well as $(1\ 3)_*F$ are $\mathfrak{S}_{[4,n]}$-invariant functions, it is possible to choose also $G \in \mathcal{O}_{X^n}^{\mathfrak{S}_{[4,n]}}$. We set

$$s(1,1,1) = s(1,3,1) = s(1,1,3) = 0 \quad , \quad s(3,1,1) = F \quad , \quad s(1,2,3) = G \, .$$

Then $s \in K_1^{\mathfrak{S}_n}$. Indeed, since $F \in \mathcal{I}_{13}$ we have

$$s(1,1,1)_{|\Delta_{13}} = s(1,3,1)_{|\Delta_{13}} = s(1,1,3)_{|\Delta_{13}} = s(3,1,1)_{|\Delta_{13}}$$

which is the condition for $s \in \ker(\varphi_1(1)) \cap \ker(\varphi_1(2))$ (see proof of lemma 3.7.10). Furthermore, $\varphi_1(s)(\{2\}; 1, 2; (3,1)) = 0$ since

$$
\begin{aligned}
s(3,2,1)_{|\Delta_{12}} = \big((1\ 3)_*s(1,2,3) \big)_{|\Delta_{12}} = (1\ 3)_* \big(s(1,2,3)_{|\Delta_{23}} \big) &= (1\ 3)_*(1\ 3)_*(F_{|\Delta_{12}}) \\
&= F_{|\Delta_{12}} = s(3,1,1)_{|\Delta_{12}} \, .
\end{aligned}
$$

Similarly, we also get that $\varphi_1(s)(\{1\}; 1, 2; (1,3)) = 0 = \varphi_1(s)(\{3\}; 1, 2; (1,3))$. Because of $\varphi_2(s)(\{1,2\}; 1, 3; 1) = f \cdot \bar{\zeta}_i$ we get the inclusion

$$\mathrm{im}(\varphi_2^{\mathfrak{S}_n}(\{1,2\}; 1, 3; 1)) \supset \mathcal{I}_{123}^{\mathfrak{S}_{\overline{(1,3)}}}\mathcal{H} \, .$$

For the other inclusion we first notice that the surjection $\mathcal{I}_{13} \to \mathcal{I}_{13}/\mathcal{I}_{13}^2$ is given under the identifications

$$\mathcal{I}_{13}/\mathcal{I}_{13} \cong N_{\Delta_{13}} \cong (\Omega_X)_{13} \cong (\mathrm{pr}_1^*\,\Omega_X)_{|\Delta_{13}} \cong (\mathrm{pr}_3^*\,\Omega_X)_{|\Delta_{13}}$$

by $s \mapsto d_1 s - d_3 s$. Here $d_i \colon \mathcal{O}_{X^n} \to \mathrm{pr}_i^*\,\Omega_X$ denotes the composition of the differential

$$d \colon \mathcal{O}_{X^n} \to \Omega_{X^n} \cong \mathrm{pr}_1^*\,\Omega_X \oplus \cdots \oplus \mathrm{pr}_n^*\,\Omega_X$$

with the projection to the i-th summand. For $s \in \mathcal{O}_{X^n}$ and $\tau = (i\ j) \in \mathfrak{S}_n$ we have $d_i(\tau_*s) = d_j s$. Let $s \in K_1^{\mathfrak{S}_n}$ with $a := s(1,1,1)$ and $b := s(1,2,3)$. Then because of $\varphi_1^{\mathfrak{S}_n}(1)(s) = 0$

$$s(1,1,3)_{|\Delta_{13}} = s(1,3,1)_{|\Delta_{13}} = s(3,1,1)_{|\Delta_{13}} = a_{|\Delta_{13}}$$

and because of $\varphi_1^{\mathfrak{S}_n}(3)(s) = 0$

$$s(1,1,3)_{|\Delta_{12}} = \big((1\ 2)_*b \big)_{|\Delta_{12}}, \; s(1,3,1)_{|\Delta_{12}} = \big((2\ 3)_*b \big)_{|\Delta_{12}}, \; s(3,1,1)_{|\Delta_{12}} = \big((1\ 3)_*b \big)_{|\Delta_{12}} \, .$$

Thus, over Δ_{123} there are the equalities

$$d_3 s(1,1,3) = d_3 b \,,\; d_3 s(1,3,1) = d_2 b \,,\; d_3 s(3,1,1) = d_1 b \,,$$
$$d_1 s(1,1,3) = (d_1 + d_3)a - d_3 b \,,\; d_1 s(1,3,1) = (d_1 + d_3)a - d_2 b \,,\; d_1 s(3,1,1) = (d_1 + d_3)a - d_1 b \,.$$

Hence, still over Δ_{123} we get

$$
\begin{aligned}
&\varphi_2(s)(\{1,2\};1,3;1) \\
={}& (d_1 - d_3)\big(s(1,1,1) - s(3,1,1) - s(1,3,1) + (1\;3)_* s(1,1,3)\big) \\
={}& d_1 a - (d_1 + d_3)a + d_1 b - (d_1 + d_3)a + d_2 b + d_3 b - d_3 a + d_1 b + d_2 b - (d_1 + d_3)a + d_3 b \\
={}& -2d_1 a - 4d_3 a + 2d_1 b + 2d_2 b + 2d_3 b \\
={}& -2(d_1 + d_2 + d_3)a + 2(d_1 + d_2 + d_3)b \\
={}& 0 \,.
\end{aligned}
$$

The fourth equality is due to the fact that a is $(2\;3)$-invariant and thus $d_2 a = d_3 a$. The last equality is due to the fact that $a_{|\Delta_{123}} = b_{|\Delta_{123}}$ and thus $(d_1 + d_2 + d_3)a = (d_1 + d_2 + d_3)b$. Now we have shown that $\varphi_2(s)(\{1,2\};1,3;1) \in \mathcal{I}_{123} \cdot \mathcal{H}$ which finishes the proof. $\qquad\square$

Corollary 3.7.13. *In the Grothendieck group* $\mathrm{K}(S^n X)$ *there is the equality*

$$\mu_! \big[E_1^{[n]} \otimes E_2^{[n]} \otimes E_3^{[n]} \big] = \big[K_0^{\mathfrak{S}_n} \big] - 3 \big[\mathcal{F}^{\mathfrak{S}_{\overline{\{1,3\}}}} \big] - \big[T_1(3)^{\mathfrak{S}_{[4,n]}} \big] + 2 \big[\mathcal{E}^{\mathfrak{S}_{[4,n]}} \big] - \big[\mathcal{H}^{\mathfrak{S}_{\overline{\{1,3\}}}} \big] + \big[\mathcal{H}_{|\Delta_{123}}^{\mathfrak{S}_{[4,n]}} \big] \,.$$

Proof. This follows by the fact that $R^i \mu_*(E_1^{[n]} \otimes E_2^{[n]} \otimes E_3^{[n]}) = 0$ for $i > 0$ (see corollary 2.5.2) and the results of this subsection. $\qquad\square$

Definition 3.7.14. We use for $m \in \mathbb{N}$ and $F^\bullet \in \mathrm{D}^b(X)$ the abbreviation (see lemma 1.3.1)

$$s^m \chi(F^\bullet) := \chi(((F^\bullet)^{\boxtimes m})^{\mathfrak{S}_m}) = \chi(S^m \mathrm{H}^*(F^\bullet)) = \binom{\chi(F^\bullet) + m - 1}{m} \,.$$

Furthermore, for $F^\bullet = \mathcal{O}_X$ we set

$$s^m \chi := s^m \chi(\mathcal{O}_X) = \chi(\mathcal{O}_{S^m X}) = \chi(S^m \mathrm{H}^*(\mathcal{O}_X)) = \binom{\chi(\mathcal{O}_X) + m - 1}{m} \,.$$

Corollary 3.7.15. *If X is projective, the Euler characteristic of the triple tensor product of tautological bundles is given by*

$$
\begin{aligned}
&\chi(E_1^{[n]} \otimes E_2^{[n]} \otimes E_3^{[n]}) \\
={}& \chi(E_1)\chi(E_2)\chi(E_3)s^{n-3}\chi
\end{aligned}
$$

$$+ \left(\chi(E_1 \otimes E_2)\chi(E_3) + \chi(E_1 \otimes E_3)\chi(E_2) + \chi(E_1 \otimes E_3)\chi(E_2) \right)\left(s^{n-2}\chi - s^{n-3}\chi\right)$$
$$+ \chi(E_1 \otimes E_2 \otimes E_3)(s^{n-1}\chi - 3s^{n-2}\chi + 2s^{n-3}\chi)$$
$$+ \chi(\Omega_X \otimes E_1 \otimes E_2 \otimes E_3)(s^{n-3}\chi - s^{n-2}\chi).$$

Proof. The first summand comes from $K_0(1,2,3)$, the second from $K_0(1,1,3)$, $K_0(1,3,1)$, $K_0(3,1,1)$, and $T_1(3)$, the third from $K_0(1,1,1)$, $\mathcal{F}$, and $\mathcal{E}$, and the fourth from $\mathcal{H}$ and $\mathcal{H}_{|\Delta_{123}}$. $\square$

3.8 Generalisations

3.8.1 Determinant line bundles

There is a homomorphism which associates to any line bundle on X its associated *determinant line bundle* on $X^{[n]}$ given by

$$\mathcal{D}\colon \operatorname{Pic} X \to \operatorname{Pic} X^{[n]} \quad , \quad L \mapsto \mathcal{D}_L := \mu^*((L^{\boxtimes n})^{\mathfrak{S}_n}).$$

Here the $\mathfrak{S}_n$-linearization of $L^{\boxtimes n}$ is given by the canonical isomorphisms $p^*_{\sigma^{-1}(i)}L \cong \sigma^* p^*_i L$, i.e. given by permutation of the factors. By [DN89, Theorem 2.3] the sheaf of invariants of $L^{\boxtimes n}$ is also the decent of $L^{\boxtimes n}$, i.e. $L^{\boxtimes n} \cong \pi^*((L^{\boxtimes n})^{\mathfrak{S}_n})$.

Remark 3.8.1. The functor $\mathcal{D}$ maps the trivial respectively the canonical line bundle to the trivial respectively the canonical line bundle, i.e. $\mathcal{D}_{\mathcal{O}_X} \cong \mathcal{O}_{X^{[n]}}$ and $\mathcal{D}_{\omega_X} \cong \omega_{X^{[n]}}$. The assertion for the trivial line bundle is true since the pull-back of the trivial line bundle along any morphism is the trivial line bundle and since taking the invariants of the trivial line bundle yields the trivial line bundle on the quotient by the group action. For a proof of $\mathcal{D}_{\omega_X} \cong \omega_{X^{[n]}}$ see [NW04, Proposition 1.6].

Lemma 3.8.2. *Let L be a line bundle on X.*

(i) For every $\mathcal{F}^\bullet \in \mathrm{D}^b(X^{[n]})$ there is a natural isomorphism $\Phi(\mathcal{F}^\bullet \otimes \mathcal{D}_L) \simeq \Phi(\mathcal{F}^\bullet) \otimes L^{\boxtimes n}$ in $\mathrm{D}_{\mathfrak{S}_n}(X^n)$.

(ii) For every $\mathcal{G}^\bullet \in \mathrm{D}^b_{\mathfrak{S}_n}(X^n)$ and every subgroup $H \leq \mathfrak{S}_n$ there is a natural isomorphism $[\pi_*(\mathcal{G}^\bullet \otimes L^{\boxtimes n})]^H \simeq (\pi_* \mathcal{G}^\bullet)^H \otimes (\pi_* L^{\boxtimes n})^{\mathfrak{S}_n}$.

(iii) For every $\mathcal{F}^\bullet \in \mathrm{D}^b(X^{[n]})$ there is a natural isomorphism

$$R\mu_*(\mathcal{F}^\bullet \otimes \mathcal{D}_L) \simeq R\mu_*\mathcal{F} \otimes (L^{\boxtimes n})^{\mathfrak{S}_n}.$$

94

Proof. By the definition of the determinant line bundle and the fact that $\pi^*(L^{\boxtimes n})^{\mathfrak{S}_n} \cong L^{\boxtimes n}$ we have

$$q^*\mathcal{D}_L \cong q^*\mu^*(L^{\boxtimes n})^{\mathfrak{S}_n} \cong p^*\pi^*(L^{\boxtimes n})^{\mathfrak{S}_n} \cong p^*L^{\boxtimes n}\,.$$

Using this, we get indeed natural isomorphisms

$$\Phi(\mathcal{F}^\bullet \otimes \mathcal{D}_L) \simeq Rp_*q^*(\mathcal{F}^\bullet \otimes \mathcal{D}_L) \simeq Rp_*\left(q^*\mathcal{F}^\bullet \otimes q^*\mathcal{D}_L\right) \simeq Rp_*\left(q^*\mathcal{F}^\bullet \otimes p^*L^{\boxtimes n}\right)$$
$$\overset{\mathrm{PF}}{\simeq} Rp_*q^*\mathcal{F}^\bullet \otimes L^{\boxtimes n}$$
$$\simeq \Phi(\mathcal{F}^\bullet) \otimes L^{\boxtimes n}\,.$$

This shows (i). For (ii) we remember that the functor $(\_)^{\mathfrak{S}_n}$ on $\mathrm{D}^b_{\mathfrak{S}_n}(X^n)$ is a abbreviation of the composition $(\_)^{\mathfrak{S}} \circ \pi_*$. Then

$$\left[\pi_*(\mathcal{G}^\bullet \otimes L^{\boxtimes n})\right]^H \simeq \left[\pi_*(\mathcal{G}^\bullet \otimes \pi^*(L^{\boxtimes n})^{\mathfrak{S}_n})\right]^H \overset{\mathrm{PF}}{\simeq} \left[\pi_*(\mathcal{G}^\bullet) \otimes (L^{\boxtimes n})^{\mathfrak{S}_n}\right]^H \overset{1.5.9}{\simeq} (\pi_*\mathcal{G}^\bullet)^H \otimes (L^{\boxtimes n})^{\mathfrak{S}_n}\,.$$

Now (iii) follows by (i),(ii) with $H = \mathfrak{S}_n$ and proposition 2.2.4 or directly by the projection formula along μ. $\qquad\square$

Corollary 3.8.3. *Let $E_1, \ldots, E_k$ be locally free sheaves and L a line bundle on X. Then $\Phi(E_1^{[n]} \otimes \cdots \otimes E_k^{[n]} \otimes \mathcal{D}_L)$ as well as $R\mu_*(E_1^{[n]} \otimes \cdots \otimes E_k^{[n]} \otimes \mathcal{D}_L)$ are cohomologically concentrated in degree zero, i.e.*

$$\Phi(E_1^{[n]} \otimes \cdots \otimes E_k^{[n]} \otimes \mathcal{D}_L) \simeq p_*q^*(E_1^{[n]} \otimes \cdots \otimes E_k^{[n]} \otimes \mathcal{D}_L)\,,$$
$$R\mu_*(E_1^{[n]} \otimes \cdots \otimes E_k^{[n]} \otimes \mathcal{D}_L) \simeq \mu_*(E_1^{[n]} \otimes \cdots \otimes E_k^{[n]} \otimes \mathcal{D}_L)\,.$$

Proof. This follows from corollary 2.5.2 and the previous lemma. $\qquad\square$

Using the above lemma and the corollary we can generalise the results of this chapter to sheaves on $X^{[n]}$ tensorised with determinant line bundles.

3.8.2 Derived functors

We can generalise the results on the the push forwards of the Grothendieck classes along the Hilbert-Chow morphism and with this also the results on the Euler characteristics from locally free sheaves on X to objects in $\mathrm{D}^b(X)$. For this we have to note that we can define the occurring functors K_0 and T_ℓ (subsection 3.1.1), H_ℓ (subsection 3.6.1), and $\mathcal{F}$, $\mathcal{E}$, and $\mathcal{H}$ (subsection 3.7.3) also on the level of derived categories.

Definition 3.8.4. Let $E_1^\bullet, \ldots, E_k^\bullet \in \mathrm{D}^b(X)$. We define for $n \in N$ derived multi-functors as follows

$$K_0(a)\colon \mathrm{D}^b(X)^k \to \mathrm{D}^b_{\mathfrak{S}_n}(X^n),$$

$$(E_1^\bullet, \ldots, E_k^\bullet) \mapsto \mathrm{pr}_{a(1)}^* E_1^\bullet \otimes^L \cdots \otimes^L \mathrm{pr}_{a(k)}^* E_k^\bullet \simeq \bigotimes_{t=1}^n \mathrm{pr}_t^*(\otimes^L_{\alpha \in a^{-1}(t)} E_\alpha^\bullet),$$

$$T_\ell(M; i, j; a)\colon \mathrm{D}^b(X)^k \to \mathrm{D}^b_{\mathfrak{S}_n}(X^n),$$

$$(E_1^\bullet, \ldots, E_k^\bullet) \mapsto \left(S^{\ell-1}\Omega_X \otimes (\otimes^L_{\alpha \in \hat{M}(a)} E_\alpha^\bullet)\right)_{ij} \otimes \bigotimes_{t=3}^n \mathrm{pr}_t^*(\otimes^L_{\alpha \in a^{-1}(t)} \mathcal{E}_\alpha^\bullet).$$

$$K_0 := \bigoplus_{a \in I_0} K_0(a) \quad, \quad T_\ell := \bigoplus_{(M; i, j; a) \in I_\ell} T_\ell(M; i, j; a)$$

$$H_\ell\colon \mathrm{D}^b(X)^k \to \mathrm{D}^b_{\mathfrak{S}_2}(X^2), \, (E_1^\bullet, \ldots, E_k^\bullet) \mapsto \Delta_*(E_1^\bullet \otimes^L \cdots \otimes^L E_k^\bullet),$$

$$\mathcal{F}\colon \mathrm{D}^b(X)^k \to \mathrm{D}^b_{\mathfrak{S}_n}(X^n) \quad (n \geq 3), \, (E_1^\bullet, E_2^\bullet, E_3^\bullet) \mapsto \left(E_1^\bullet \otimes^L E_2^\bullet \otimes^L E_3^\bullet\right)_{13},$$

$$\mathcal{E}\colon \mathrm{D}^b(X)^k \to \mathrm{D}^b_{\mathfrak{S}_n}(X^n) \quad (n \geq 3), \, (E_1^\bullet, E_2^\bullet, E_3^\bullet) \mapsto \left(E_1^\bullet \otimes^L E_2^\bullet \otimes^L E_3^\bullet\right)_{123},$$

$$\mathcal{H}\colon \mathrm{D}^b(X)^k \to \mathrm{D}^b_{\mathfrak{S}_n}(X^n) \quad (n \geq 3), \, (E_1^\bullet, E_2^\bullet, E_3^\bullet) \mapsto \left(\Omega_X \otimes E_1^\bullet \otimes^L E_2^\bullet \otimes^L E_3^\bullet\right)_{13}$$

$$\mathcal{H}_{123}\colon \mathrm{D}^b(X)^k \to \mathrm{D}^b_{\mathfrak{S}_n}(X^n) \quad (n \geq 3), \, (E_1^\bullet, E_2^\bullet, E_3^\bullet) \mapsto \left(\Omega_X \otimes E_1^\bullet \otimes^L E_2^\bullet \otimes^L E_3^\bullet\right)_{123}$$

The empty derived tensor product has to be interpreted as the sheaf $\mathcal{O}_X$.

The functor $(\_)_I$ is the composition of the pull-back along the flat morphism p_I and the closed embedding ι_I. Thus, it is exact. The only derived functor occurring in the above multi-functors is the tensor product on X. Thus the images under the functors can be computed by replacing $E_1^\bullet, \ldots, E_k^\bullet$ by locally free resolutions. In particular, for $E_1, \ldots, E_k$ locally free sheaves on X the functors K_0, T_ℓ, H_ℓ, $\mathcal{E}$, and $\mathcal{H}$ coincide with the functors defined before and $\mathcal{H}_{123}$ coincides with $\mathcal{H}_{|\Delta_{123}}$. Again we will often drop the arguments of the functors in the notations. We also define as in subsection 3.7.3 the functors $T_1(1)$ and $T_1(3)$ as the direct sum of the $T_1(M; i, j; a)$ in the case $k = 3$ over $\tilde{J}^1$ respectively $\tilde{J}^3$. Again $T_1(1)^{\mathfrak{S}_{\overline{\{1,3\}}}} \cong (\mathcal{F}^{\mathfrak{S}_{\overline{\{1,3\}}}})^{\oplus 3}$.

3.8.3 Generalised results

Theorem 3.8.5. *Let $E_1, \ldots, E_k$ be locally free sheaves and L a line bundle on X. Then on X^n there is the equality*

$$K_k \otimes L^{\boxtimes n} = p_* q^*(E_1^{[n]} \otimes \cdots \otimes E_k^{[n]} \otimes \mathcal{D}_L)$$

of subsheaves of $K_0 \otimes L^{\boxtimes n}$. Also, for every $\ell = 1, \ldots, k$ we have $K_\ell \otimes L^{\boxtimes n} = \ker(\varphi_\ell \otimes \mathrm{id}_{L^{\boxtimes n}})$.

Proof. Use theorem 3.1.7 and lemma 3.8.2. The second assertion is due to the fact that tensoring with the line bundle $L^{\boxtimes n}$ is exact. $\qquad\square$

Proposition 3.8.6. *Let $E_1, \ldots, E_k$ be locally free sheaves and L a line bundle on X. Then*

$$\mathrm{H}^{2n}(X^{[n]}, E_1^{[n]} \otimes \cdots \otimes E_k^{[n]} \otimes \mathcal{D}_L) \cong \mathrm{H}^{2n}(X^n, K_0 \otimes L^{\boxtimes n})^{\mathfrak{S}_n}$$

$$\cong \bigoplus_{a \in J_0(1)} \bigotimes_{r=1}^{\max a} \mathrm{H}^2 \Big(\bigotimes_{t \in a^{-1}(r)} E_t \otimes L \Big) \otimes S^{n - \max a} \, \mathrm{H}^2(L) \, .$$

Proof. This follows from theorem 3.8.5 the same way proposition 3.4.5 followed from theorem 3.1.7. □

Theorem 3.8.7. *Let $E_1, \ldots, E_k$ be locally free sheaves and L a line bundle on X. Then on X^2 for every $\ell = 1, \ldots, k$ the sequences*

$$0 \to K_\ell \otimes L^{\boxtimes n} \to K_{\ell-1} \otimes L^{\boxtimes n} \to T_\ell^0 \otimes L^{\boxtimes n} \to T_\ell^1 \otimes L^{\boxtimes n} \to \cdots \to T_\ell^{k-\ell} \otimes L^{\boxtimes n} \to 0$$

are exact.

Proof. We tensorize the exact sequences of 3.6.6 with the line bundle $L^{\boxtimes n}$. □

Proposition 3.8.8. *On $S^2 X$ there are for $\ell = 1, \ldots, k$ exact sequences*

$$0 \to (K_\ell \otimes L^{\boxtimes n})^{\mathfrak{S}_2} \to (K_{\ell-1} \otimes L^{\boxtimes n})^{\mathfrak{S}_2} \to \pi_*(H_\ell \otimes L^{\boxtimes n})^{\oplus N(k,\ell)} \to 0 \, .$$

In particular $(K_k \otimes L^{\boxtimes n})^{\mathfrak{S}_2} = (K_{k-1} \otimes L^{\boxtimes n})^{\mathfrak{S}_2}$.

Proof. We tensorise the exact sequences of proposition 3.6.9 with the line bundle $(L^{\boxtimes 2})^{\mathfrak{S}_2}$ and use lemma 3.8.2 (ii). □

Lemma 3.8.9. *For $E_1, \ldots, E_k$ locally free sheaves and L a line bundle on X there is in the Grothendieck group $\mathrm{K}(S^2 X)$ the equality*

$$\mu_! \big[(E_1^{[2]} \otimes \cdots \otimes E_k^{[2]} \otimes \mathcal{D}_L) \big] = \big[(K_0 \otimes L^{\boxtimes n})^{\mathfrak{S}_2} \big] - \sum_{\ell=1}^k N(k, \ell) \big[(H_\ell \otimes L^{\boxtimes n})^{\mathfrak{S}_2} \big] \, .$$

Proof. We multiply both sides of the equation in corollary 3.6.10 via the $\mathrm{K}^0(S^2 X)$-module structure (see section 1.3) of $\mathrm{K}(S^2 X)$ with $[(L^{\boxtimes n})^{\mathfrak{S}_n}]$. Then we apply 3.8.2 (iii) to the left-hand side and 3.8.2 (ii) to the right-hand side. □

Proposition 3.8.10. *For $E_1^\bullet, \ldots, E_k^\bullet \in \mathrm{D}^b(X)$ and L a line bundle on X there is in the Grothendieck group $\mathrm{K}(S^2 X)$ the equality*

$$\mu_! \big[((E_1^\bullet)^{[2]} \otimes \cdots \otimes (E_k^\bullet)^{[2]} \otimes \mathcal{D}_L) \big] = \big[(K_0 \otimes L^{\boxtimes n})^{\mathfrak{S}_2} \big] - \sum_{\ell=1}^k N(k, \ell) \big[(H_\ell \otimes L^{\boxtimes n})^{\mathfrak{S}_2} \big] \, .$$

Proof. We denote by A the left-hand side of the equation and by B the right hand side, both as functions in $E_1^\bullet, \ldots, E_k^\bullet$. We choose locally free resolutions $F_i^\bullet$ of $E_i^\bullet$ for $i \in [k]$. Since the functor $(\_)^{[n]}$ is exact and $R\mu_*$ coincides with μ_* on tensor products of tautological bundles and determinant line bundles by corollary 3.8.3 we have for the left-hand side

$$A((E_i^\bullet)_i) = A((F_i^\bullet)_i) = \sum_{q \in \mathbb{Z}} (-1)^q \left(\sum_{p_1 + \cdots + p_k = q} A(F_1^{q_1}, \ldots, F_k^{q_k}) \right).$$

Since the derived functors occuring on the right-hand side coincide with their non-derived versions on locally free sheaves, we get also

$$B((E_i^\bullet)_i) = B((F_i^\bullet)_i) = \sum_{q \in \mathbb{Z}} (-1)^q \left(\sum_{p_1 + \cdots + p_k = q} B(F_1^{q_1}, \ldots, F_k^{q_k}) \right)$$

and the assertion follows by the previous lemma. $\qquad\square$

Theorem 3.8.11. *Let X be a smooth projective surface. Let $E_1^\bullet, \ldots, E_k^\bullet \in D^b(X)$ and L a line bundle on X. Then there is the formula*

$$\chi_{X^{[2]}} \left((E_1^\bullet)^{[2]} \otimes^L \cdots \otimes^L (E_k^\bullet)^{[2]} \otimes^L \mathcal{D}_L \right)$$
$$= \sum_{\{P_1, P_2\}} \chi \left(\otimes_{t \in P_1}^L E_t \otimes L \right) \cdot \chi \left(\otimes_{t \in P_2}^L E_t \otimes L \right) - \sum_{\ell=1}^{k-1} N(k, \ell) \cdot \chi \left(\otimes_{t \in [k]}^L E_t \otimes S^{\ell-1} \Omega_X \otimes L^{\otimes 2} \right).$$

The sum is taken over all partitions $\{P_1, P_2\}$ of the set $[k]$ of length at most two.

Proof. This follows from the previous proposition the same way proposition 3.6.16 followed from corollary 3.6.10. $\qquad\square$

Theorem 3.8.12. *Let $E_1, \ldots, E_k$ be locally free sheaves and L a line bundle on X and let $n \geq 3$. Then the sheaf*

$$\mu_*(E_1^{[n]} \otimes \cdots \otimes E_k^{[n]} \otimes \mathcal{D}_L) \cong (K_2 \otimes L^{\boxtimes n})^{\mathfrak{S}_n}$$

is given by the following exact sequences

$$0 \to \ker(\varphi_1^{\mathfrak{S}_n}(1)) \otimes (L^{\boxtimes n})^{\mathfrak{S}_n} \to (K_0 \otimes L^{\boxtimes n})^{\mathfrak{S}_n} \to (T_1(1) \otimes L^{\boxtimes n})^{\mathfrak{S}_{\overline{\{1,3\}}}} \to 0,$$
$$0 \to (K_1 \otimes L^{\boxtimes n})^{\mathfrak{S}_n} \to \ker(\varphi_1^{\mathfrak{S}_n}(1)) \otimes (L^{\boxtimes n})^{\mathfrak{S}_n} \to (T_1(3) \otimes L^{\boxtimes n})^{\mathfrak{S}_{[4,n]}} \to (\mathcal{E}^2 \otimes L^{\boxtimes n})^{\mathfrak{S}_{[4,n]}} \to 0,$$
$$0 \to (K_2 \otimes L^{\boxtimes n})^{\mathfrak{S}_n} \to (K_1 \otimes L^{\boxtimes n})^{\mathfrak{S}_n} \to (\mathcal{H} \otimes L^{\boxtimes n})^{\mathfrak{S}_{\overline{\{1,3\}}}} \to (\mathcal{H}_{|\Delta_{123}} \otimes L^{\boxtimes n})^{\mathfrak{S}_{[4,n]}} \to 0.$$

Proof. The isomorphism $\mu_*(E_1^{[n]} \otimes \cdots \otimes E_k^{[n]} \otimes \mathcal{D}_L) \cong (K_2 \otimes L^{\boxtimes n})^{\mathfrak{S}_n}$ follows by 3.8.2 (iii) and 3.2.4. We get the exactness of the sequences by tensoring the exact sequences of 3.7.10, 3.7.11, and 3.7.12 by $(L^{\boxtimes n})^{\mathfrak{S}_n}$ and using lemma 3.8.2 (ii). $\qquad\square$

Proposition 3.8.13. *Let* $E_1^\bullet, E_2^\bullet, E_3^\bullet \in D^b(X)$ *and* L *a line bundle on* X. *Then for* $n \geq 3$ *there is in the Grothendieck group* $K(S^n X)$ *the equality*

$$\mu_! \big[(E_1^\bullet)^{[n]} \otimes^L (E_2^\bullet)^{[n]} \otimes^L (E_3^\bullet)^{[n]} \otimes \mathcal{D}_L \big]$$
$$= \big[(K_0 \otimes L^{\boxtimes n})^{\mathfrak{S}_n} \big] - 3\big[(\mathcal{F} \otimes L^{\boxtimes n})^{\mathfrak{S}_{\overline{\{1,3\}}}} \big] - \big[(T_1(3) \otimes L^{\boxtimes n})^{\mathfrak{S}_{[4,n]}} \big]$$
$$+ 2\big[(\mathcal{E} \otimes L^{\boxtimes n})^{\mathfrak{S}_{[4,n]}} \big] - \big[(\mathcal{H} \otimes L^{\boxtimes n})^{\mathfrak{S}_{\overline{\{1,3\}}}} \big] + \big[(\mathcal{H}_{123} \otimes L^{\boxtimes n})^{\mathfrak{S}_{[4,n]}} \big].$$

Proof. This follows from 3.7.13 the same way 3.8.10 followed from 3.6.10. $\qquad\square$

We use the notation of 3.7.14, namely

$$s^m \chi(L) := \chi((L^{\boxtimes m})^{\mathfrak{S}_m}) = \chi(S^m \mathrm{H}^\bullet(L)) = \binom{\chi(L) + m - 1}{m}.$$

Theorem 3.8.14. *Let* $E_1^\bullet, E_2^\bullet, E_3^\bullet \in D^b(X)$ *and* L *a line bundle on a smooth projective surface* X. *Then for* $n \geq 3$ *there is the formula*

$$\chi((E_1^\bullet)^{[n]} \otimes^L (E_2^\bullet)^{[n]} \otimes^L (E_3^\bullet)^{[n]} \otimes \mathcal{D}_L)$$
$$= \chi(E_1^\bullet \otimes L)\chi(E_2^\bullet \otimes L)\chi(E_3^\bullet \otimes L)s^{n-3}\chi(L)$$
$$+ \Big(\sum_I \chi(E_a^\bullet \otimes^L E_b^\bullet \otimes L)\chi(E_c^\bullet \otimes L) \Big) \cdot s^{n-2}\chi(L)$$
$$- \Big(\sum_I \chi(E_a^\bullet \otimes^L E_b^\bullet \otimes L \otimes L)\chi(E_c^\bullet \otimes L) \Big) \cdot s^{n-3}\chi(L)$$
$$+ \chi(E_1^\bullet \otimes^L E_2^\bullet \otimes^L E_3^\bullet \otimes L)s^{n-1}\chi(L)$$
$$- 3\chi(E_1^\bullet \otimes^L E_2^\bullet \otimes^L E_3^\bullet \otimes L \otimes L)s^{n-2}\chi(L)$$
$$+ 2\chi(E_1^\bullet \otimes^L E_2^\bullet \otimes^L E_3^\bullet \otimes L \otimes L \otimes L)s^{n-3}\chi(L)$$
$$- \chi(\Omega_X \otimes E_1^\bullet \otimes^L E_2^\bullet \otimes^L E_3^\bullet \otimes L \otimes L)s^{n-2}\chi(L)$$
$$+ \chi(\Omega_X \otimes E_1^\bullet \otimes^L E_2^\bullet \otimes^L E_3^\bullet \otimes L \otimes L \otimes L)s^{n-3}\chi(L).$$

Here I *denotes the index set*

$$I = \{(a = 1, b = 2, c = 3), (a = 1, b = 3, c = 2), (a = 2, b = 3, c = 1)\}.$$

Proof. This follows from the previous proposition the same way 3.7.15 followed from 3.7.13. $\qquad\square$

Chapter 4

Extension groups of twisted tautological objects

Throughout the rest of the text let X be a smooth quasi-projective surface over the complex numbers $\mathbb{C}$, $n \geq 2$ a positive integer and $X^{[n]}$ the Hilbert scheme of n points on X. We will use the Bridgeland-King-Reid equivalence to compute the extension groups of certain objects $\mathcal{E}^\bullet, \mathcal{F}^\bullet \in D^b(X^{[n]})$ by the formula (see corollary 2.2.3)

$$\mathrm{Ext}^i_{X^{[n]}}(\mathcal{E}^\bullet, \mathcal{F}^\bullet) \cong \mathfrak{S}_n \, \mathrm{Ext}^i_{X^n}(\Phi(\mathcal{E}^\bullet), \Phi(\mathcal{F}^\bullet)).$$

We can rewrite the right hand side as

$$\mathfrak{S}_n \, \mathrm{Ext}^i_{X^n}(\Phi(\mathcal{E}^\bullet), \Phi(\mathcal{F}^\bullet))) \cong R^i \, \Gamma(S^n X, [\pi_* R \, \mathcal{H}om_{X^n}(\Phi(\mathcal{E}^\bullet), \Phi(\mathcal{F}^\bullet)]^{\mathfrak{S}_n})$$
$$= \mathrm{H}^i(S^n X, [\pi_* R \, \mathcal{H}om_{X^n}(\Phi(\mathcal{E}^\bullet), \Phi(\mathcal{E}^\bullet)]^{\mathfrak{S}_n}).$$

The functors π_* and $[\_]^{\mathfrak{S}_n}$ are indeed exact and hence need not be derived. We will first compute the inner expression $[\pi_* R \, \mathcal{H}om_{X^n}(\Phi(\mathcal{E}^\bullet), \Phi(\mathcal{E}^\bullet))]^{\mathfrak{S}_n}$. We abbreviate the occurring bifunctor by

$$\underline{\mathrm{Hom}}(\_,\_) := [\pi_* \, \mathcal{H}om_{X^n}(\_,\_)]^{\mathfrak{S}_n}$$

and set $\underline{\mathrm{Ext}}^i = R^i \, \underline{\mathrm{Hom}}$. Because of the exactness of $[\_]^{\mathfrak{S}_n} := [\_]^{\mathfrak{S}_n} \circ \pi_*$ we have

$$R \, \underline{\mathrm{Hom}}(\_,\_) \simeq [R \, \mathcal{H}om_{X^n}(\_,\_)]^{\mathfrak{S}_n}.$$

4.1 The case of tautological bundles

4.1.1 Computation of the <u>Hom</u>s

We denote by $\mathbb{D}$ the big diagonal in X^n, i.e. $\mathbb{D} = \{(x_1, \ldots, x_n) \mid \#\{x_1, \ldots, x_n\} < n\} \subset X^n$, and by $U = X^n \setminus \mathbb{D}$ its open complement in X^n.

Proposition 4.1.1. *Let* $k \in \mathbb{N}$ *and* $E_1, \ldots, E_k, F \in \mathrm{Coh}(X)$ *be locally free sheaves. Then there are natural isomorphisms*

(i) $\underline{\mathrm{Hom}}(\Phi(E_1^{[n]} \otimes \cdots \otimes E_k^{[n]}), \mathcal{O}_{X^n}) \simeq \underline{\mathrm{Hom}}(C_{E_1}^0 \otimes \cdots \otimes C_{E_k}^0, \mathcal{O}_{X^n}),$

(ii) $\underline{\mathrm{Hom}}(\Phi(E_1^{[n]} \otimes \cdots \otimes E_k^{[n]}), \Phi(F^{[n]})) \simeq \underline{\mathrm{Hom}}(C_{E_1}^0 \otimes \cdots \otimes C_{E_k}^0, C_F^0).$

Proof. By theorem 2.7.1 we have $p_* q^*(E_1^{[n]} \otimes \cdots \otimes E_k^{[n]}) \simeq \Phi(E_1^{[n]} \otimes \cdots \otimes E_k^{[n]})$. By theorem 3.1.7 $p_* q^*(\otimes_{i=1}^k E_i^{[n]})$ can be identified with the subsheaf K_k of $C_{E_1}^0 \otimes \cdots \otimes C_{E_k}^0$. Since the T_ℓ are all supported on $\mathbb{D}$, we have

$$p_* q^*(E_1^{[n]} \otimes \cdots \otimes E_k^{[n]})_{|U} = (C_{E_1}^0 \otimes \cdots \otimes C_{E_k}^0)_{|U}.$$

Since X^n is normal and $\mathbb{D}$ of codimension 2, lemma 1.5.28 yields the first isomorphism of the proposition (even before taking invariants). For the second one we first have to show that for every open $W \subset S^n X$ every $\mathfrak{S}_n$-equivariant morphism

$$\varphi \colon p_* q^*(E_1^{[n]} \otimes \cdots \otimes E_k^{[n]})_{|\pi^{-1}W} \to C_{F|\pi^{-1}W}^0$$

factorizes over $p_* q^* F_{|\pi^{-1}W}^{[n]}$ which will yield

$$\underline{\mathrm{Hom}}(p_* q^*(E_1^{[n]} \otimes \cdots \otimes E_k^{[n]}), p_* q^* F^{[n]}) \cong \underline{\mathrm{Hom}}(p_* q^*(E_1^{[n]} \otimes \cdots \otimes E_k^{[n]}), C_F^0).$$

Since X is quasi-projective $S^n X$ has an open affine covering consisting of subsets of the form $S^n U$ with $U \subset X$ open and affine (see lemma 3.1.5). Thus, it suffices to consider $W = S^n U$ and hence $\pi^{-1} W = U^n$ for an open affine $U \subset X$. Recall (theorem 2.5.3) that $p_* q^* F^{[n]} = \ker(d_F^0 \colon C_F^0 \to C_F^1)$. Thus, we can describe the space of sections of $p_* q^* F^{[n]}$ over U^n as the sections $s = (s_1, \ldots, s_n)$ of C_F^0 with $s_{i|\Delta_{ij}} = s_{j|\Delta_{ij}}$ for every distinct $i, j \in [n]$ or alternatively

$$p_* q^* F^{[n]}(U^n) = \{s \in C_F^0(U^n) \mid s_{|\Delta_{ij}} \text{ is } (i\ j)\text{-invariant } \forall i, j \in [n], i \neq j\}.$$

By remark 3.3.1 we also have for the sections over U^n the inclusions

$$p_* q^*(\otimes_{i=1}^n E_i^{[n]}) \subset K_1 \subset \{s \in C_{E_1}^0 \otimes \cdots \otimes C_{E_k}^0 \mid s_{|\Delta_{ij}} \text{ is } (i\ j)\text{-invariant } \forall i, j \in [n], i \neq j\}.$$

For an $\mathfrak{S}_n$-equivariant morphism $\varphi\colon p_*q^*(\otimes_i E_i^{[n]})_{|U^n} \to C^0_{F|U^n}$ this gives

$$(i\ j)(\varphi(s))_{|\Delta_{ij}} = \varphi((i\ j)s_{|\Delta_{ij}}) = \varphi(s_{|\Delta_{ij}}) = \varphi(s)_{|\Delta_{ij}}\,.$$

Thus, $\varphi(s) \in p_*q^*F^{[n]}$ as desired. Now we can again apply lemma 1.5.28 and get

$$\underline{\mathrm{Hom}}(p_*q^*(E_1^{[n]} \otimes \cdots \otimes E_k^{[n]}), p_*q^*F^{[n]}) \cong \underline{\mathrm{Hom}}(p_*q^*(E_1^{[n]} \otimes \cdots \otimes E_k^{[n]}), C^0_F)$$
$$\cong \underline{\mathrm{Hom}}(C^0_{E_1} \otimes \cdots \otimes C^0_{E_k}, C^0_F)\,.$$

$\square$

4.1.2 Vanishing of the higher $\underline{\mathrm{Ext}}^i(\Phi(F^{[n]}), \mathcal{O}_{X^n})$

Remark 4.1.2. Let $H \leq \mathfrak{S}_n$ be a subgroup, E a H-sheaf and F a $\mathfrak{S}_n$-sheaf on X^n. Since X^n has a covering by open affine $\mathfrak{S}_n$-invariant subsets, namely by U^n for $U \subset X$ open and affine (lemma 3.7.1), the adjoint property of the inflation functor globalises to a natural isomorphism

$$[\mathcal{H}om(\mathrm{Inf}_H^{\mathfrak{S}_n} E, F)]^{\mathfrak{S}_n} \cong [\mathcal{H}om(E, F)]^H\,.$$

This also gives a formula for the derived functors $R\mathcal{H}om$, namely

$$[\mathcal{H}om(\mathrm{Inf}_H^{\mathfrak{S}_n} E^\bullet, F^\bullet)]^{\mathfrak{S}_n} \simeq [\mathcal{H}om(E^\bullet, F^\bullet)]^H$$

for $E^\bullet \in \mathrm{D}_H^b(X^n)$ and $F^\bullet \in \mathrm{D}_{\mathfrak{S}_n}^b(X^n)$. Alternatively, we can apply lemma 1.5.6 with regard to the functor $\mathcal{H}om(\_, F)$ to get the same isomorphism.

Lemma 4.1.3. Let F be a locally free sheaf on X. Then $[(C_F^p)^\vee]^{\mathfrak{S}_n} \simeq 0$ for $p > 0$.

Proof. Using that $C_F^p \cong \mathrm{Inf}_{\mathfrak{S}_{[p+1]} \times \mathfrak{S}_{\overline{[p+1]}}}^{\mathfrak{S}_n} F_{[p+1]}$ (remark 2.4.3) and the previous remark we get $[(C_F^p)^\vee]^{\mathfrak{S}_n} \simeq [F_{[p+1]}^\vee]^{\mathfrak{S}_{[p+1]} \times \mathfrak{S}_{\overline{[p+1]}}}$. By remark 2.4.2, the equivariant Grothendieck duality for regular closed embeddings (proposition 1.4.9), and the fact that F is locally free, we have in $\mathrm{D}_{\mathfrak{S}_{[p+1]} \times \mathfrak{S}_{\overline{[p+1]}}}^b(X^n)$ isomorphisms

$$(F_{[p+1]})^\vee \simeq R\mathcal{H}om_{X^n}\left(\iota_{[p+1]*}(\mathfrak{a} \otimes p_{[p+1]}^*F), \mathcal{O}_{X^n}\right)$$
$$\simeq \iota_{[p+1]*}R\mathcal{H}om_{\Delta_{[p+1]}}(\mathfrak{a} \otimes p_{[p+1]}^*F, \iota_{[p+1]}^!\mathcal{O}_{X^n})$$
$$\simeq \iota_{[p+1]*}\left(\mathfrak{a} \otimes (p_{[p+1]}^*F)^\vee \otimes (\wedge^{2p}\iota_{[p+1]}^*\mathcal{I}_{[p+1]})^\vee\right)[-2p]\,.$$

The group $\mathfrak{S}_{[p+1]}$ acts trivially on $p_{[p+1]}^*F$ and on $\wedge^{2p}\iota_{[p+1]}^*\mathcal{I}_I$ (lemma 1.5.21). Thus, it acts alternating on the whole $\mathfrak{a} \otimes (p_{[p+1]}^*F)^\vee \otimes (\wedge^{2p}\iota_{[p+1]}^*\mathcal{I}_I)^\vee$ which makes the $\mathfrak{S}_{[p+1]}$-invariants vanish. Clearly, this implies the vanishing of the $\mathfrak{S}_{[p+1]} \times \mathfrak{S}_{\overline{[p+1]}}$-invariants. $\square$

Proposition 4.1.4. *For every locally free sheaf F on X there is a natural isomorphism* $[(\Phi(F^{[n]}))^\vee]^{\mathfrak{S}_n} \simeq [(C_F^0)^\vee]^{\mathfrak{S}_n}$ *in* $D^b(S^n X)$. *In particular, all* $\underline{\mathrm{Ext}}^i(\Phi(F^{[n]}), \mathcal{O}_{X^n})$ *vanish for* $i > 0$.

Proof. By theorem 2.5.3 in $D^b_{\mathfrak{S}_n}(X^n)$ there is the isomorphism $\Phi(F^{[n]}) \simeq C_F^\bullet$. The $\mathfrak{S}_n$-sheaf C_F^0 is locally free, hence $[(\_)^\vee]^{\mathfrak{S}_n}$-acyclic. For $p > 0$ the terms C_F^p are also $[(\_)^\vee]^{\mathfrak{S}_n}$-acyclic by the previous lemma. Thus, $[(\Phi(F^{[n]}))^\vee]^{\mathfrak{S}_n}$ can be computed by applying the non derived functor $[(\_)^\vee]^{\mathfrak{S}_n}$ to $C_F^\bullet$. Again by the previous lemma $[(C_F^p)^\vee]^{\mathfrak{S}_n} = 0$ for $p > 0$ and the proposition follows. $\qquad\square$

4.1.3 Vanishing of the higher $\underline{\mathrm{Ext}}^i(\Phi(E^{[n]}), \Phi(F^{[n]}))$

Remark 4.1.5. For $K \subset M \subset [n]$ we denote the closed embedding of the corresponding partial diagonals by $\iota_M^K \colon \Delta_M \to \Delta_K$. For an subset $I \subset [n]$ with $|I| \geq 2$ the codimension of the partial diagonal $\Delta_I = \{(x_1, \ldots, x_n) \mid x_i = x_j \,\forall i, j \in I\}$ in X^n is $2(|I| - 1)$. For $|I| \leq 1$ we set $\Delta_I := X^n$. Let I and J be subsets of $[n]$. If $I \cap J = \emptyset$ the partial diagonals Δ_I and Δ_J intersect properly in $X^n = \Delta_{I \cap J}$. In this case we denote the embeddings of $\Delta_I \cap \Delta_J$ into Δ_I respectively Δ_J by $\iota_{I \cup J}^I$ respectively $\iota_{I \cup J}^J$ although the intersection does not equal $\Delta_{I \cup J}$. If $I \cap J \neq \emptyset$ we have indeed $\Delta_I \cap \Delta_J = \Delta_{I \cup J}$. Furthermore,

$$\mathrm{codim}(\Delta_{I \cup J}, \Delta_{I \cap J}) = 2|(I \cup J) \setminus (I \cap J)| = 2|I \setminus (I \cap J)| + 2|J \setminus (I \cap J)|$$
$$= \mathrm{codim}(\Delta_I, \Delta_{I \cap J}) + \mathrm{codim}(\Delta_J, \Delta_{I \cap J}),$$

i.e. Δ_I and Δ_J again intersect properly in $\Delta_{I \cap J}$. In summary, for $I, J \subset [n]$ the diagram

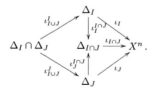

fulfils the assumptions of lemma 1.5.17. This yields for every Cohen-Macauley sheaf $\mathcal{F}$ on Δ_J and any $q \in \mathbb{Z}$ the formula

$$L^{-q} \iota_I^* \iota_{J*} \mathcal{F} \cong \iota_{I \cup J*}^I \iota_{I \cup J}^{J*} \left(\mathcal{F} \otimes (\iota_J^{I \cap J*} \wedge^q N_{I \cap J}^\vee) \right).$$

Let E and F be locally free sheaves on X. In order to compute the invariants of the higher extension sheaves we will use the spectral sequence A associated to the functor

$\underline{\mathrm{Hom}}(\_, p_*q^*F^{[n]})$ given by (see e.g. [Huy06, Remark 2.67])

$$A_1^{p,q} = [\mathcal{E}xt^q(C_E^{-p}, p_*q^*F^{[n]})]^{\mathfrak{S}_n} \implies A^m = [\mathcal{E}xt^m(C_E^{\bullet}, p_*q^*F^{[n]})]^{\mathfrak{S}_n} \cong [\mathcal{E}xt^m(\Phi(E^{[n]}), \Phi(F^{[n]})]^{\mathfrak{S}_n}.$$

The terms in the k-th column of A_1 in turn are computed by the spectral sequence $B(k)$ associated to $\underline{\mathrm{Hom}}(C_E^k, \_)$ and given by

$$B(k)_1^{p,q} = [\mathcal{E}xt^q(C_E^k, C_F^p)]^{\mathfrak{S}_n} \implies B(k)^m = [\mathcal{E}xt^m(C_E^k, C_F^{\bullet})]^{\mathfrak{S}_n} \simeq [\mathcal{E}xt^m(C_E^k, p_*q^*F^{[n]})]^{\mathfrak{S}_n}.$$

Here as direct summands terms of the form $\mathcal{E}xt^q(E_I, F_J)$ occur. For $I, J \in [n]$ with $\#I = c+1$ and $\#J = d+1$ these are $\mathfrak{S}_{I,J}$-equivariant sheaves where

$$\mathfrak{S}_{I,J} := \mathrm{Stab}_{\mathfrak{S}_n}(I, J) = (\mathfrak{S}_I \times \mathfrak{S}_{\bar{I}}) \cap (\mathfrak{S}_J \times \mathfrak{S}_{\bar{J}}) = \mathfrak{S}_{I\setminus J} \times \mathfrak{S}_{J\setminus I} \times \mathfrak{S}_{I\cap J} \times \mathfrak{S}_{\overline{I\cup J}}.$$

In $\mathrm{D}^b_{\mathfrak{S}_{I,J}}(X^n)$ we have the isomorphisms

$$\begin{aligned} R\mathcal{H}om_{X^n}(E_I, F_J) &\simeq E_I^{\vee} \otimes^L F_J \\ &\overset{1.4.9}{\simeq} \iota_{I*}\left(p_I^*E^{\vee} \otimes \mathfrak{a}_I \otimes (\wedge^{2c}\iota_I^*\mathcal{I}_I)^{\vee}\right) \otimes^L_{X^n} \iota_{J*}(p_J^*F \otimes \mathfrak{a}_J)[-2c] \\ &\overset{\mathrm{PF}}{\simeq} \iota_{I*}\left(p_I^*E^{\vee} \otimes \mathfrak{a}_I \otimes (\wedge^{2c}\iota_I^*\mathcal{I}_I)^{\vee} \otimes L\iota_I^*\iota_{J*}(p_J^*F \otimes \mathfrak{a}_J)\right)[-2c]. \end{aligned}$$

We see that $R\mathcal{H}om_{X^n}(E_I, F_J)$ has no cohomology in degrees greater than $2c$. Taking the $(2c-q)$-th cohomology for $q \geq 0$ on both sides yields

$$\begin{aligned} &\mathcal{E}xt^{2c-q}(E_I, F_J) \\ &\overset{\mathrm{lf}}{\cong} \iota_{I*}\left(p_I^*E^{\vee} \otimes \mathfrak{a}_I \otimes (\wedge^{2c}\iota_I^*\mathcal{I}_I)^{\vee} \otimes L^{-q}\iota_I^*\iota_{J*}(p_J^*F \otimes \mathfrak{a}_J)\right) \\ &\overset{4.1.5}{\cong} \iota_{I*}\left((p_I^*E \otimes \mathfrak{a}_I \otimes (\wedge^{2c}\iota_I^*\mathcal{I}_I))^{\vee} \otimes \iota_{I\cup J*}^I(\wedge^q N^{\vee}_{I\cap J|\Delta_I \cap \Delta_J} \otimes (p_J^*F \otimes \mathfrak{a}_J)_{|\Delta_I \cap \Delta_J})\right) \\ &\overset{\mathrm{PF}}{\cong} \iota_{I\cup J*}\underbrace{\left(\iota_{I\cup J}^{I*}(p_I^*E \otimes \mathfrak{a}_I \otimes (\wedge^{2c}\iota_I^*\mathcal{I}_I))^{\vee} \otimes \wedge^q N^{\vee}_{I\cap J|\Delta_I \cap \Delta_J} \otimes (p_J^*F \otimes \mathfrak{a}_J)_{|\Delta_I \cap \Delta_J}\right)}. \end{aligned}$$

We abbreviate the underbraced $\mathfrak{S}_{I,J}$-sheaf on $\Delta_I \cap \Delta_J$ by $T_{I,J}^q$. We have $\mathcal{E}xt^{2c-q}(E_I, F_J) = 0$ for $q > \mathrm{codim}(\Delta_{I\cup J}, X^n) = 2(|I \cap J| - 1)$.

Lemma 4.1.6. *Let $I, J \subset [n]$ such that $\#(I \setminus J) \geq 2$ or $\#(J \setminus I) \geq 2$. Then*

$$[R\mathcal{H}om_{X^n}(E_I, F_J)]^{\mathfrak{S}_{I,J}} \simeq 0.$$

Proof. Since $\mathcal{E}xt^{2c-q}(E_I, F_J)$ is by the computation above the push-forward of the sheaf $T_{I,j}^q$ which is defined on $\Delta_I \cap \Delta_J$, the $\mathfrak{S}_{I\setminus J} \times \mathfrak{S}_{J\setminus I} \times \mathfrak{S}_{I\cap J}$-linearization of it is just an ordinary group action. Assume that $\#(I \setminus J) \geq 2$ and choose a transposition $\tau \in \mathfrak{S}_{I\setminus J}$. Then τ acts by

-1 on $\mathfrak{a}_I$ and trivially on all other tensor factors of $T^q_{I,J}$. Hence, it acts by -1 on the whole $T^q_{I,J}$ which makes the invariants vanish. The case $\#(J \setminus I) \geq 2$ is analogous. □

Remark 4.1.7. For $I \cap J \neq \emptyset$ the $\mathfrak{S}_{I \cap J}$-action on $T^q_{I,J}$ is given by the $\mathfrak{S}_{I \cap J}$-action on the factor $\wedge^q N^\vee_{I \cap J | \Delta_{I \cup J}}$ since $\mathfrak{S}_{I \cap J}$ acts alternating on two of the other tensor factors of $T^q_{I,J}$, namely on $\mathfrak{a}_I$ and $\mathfrak{a}_J$, and trivially on the remaining three. Thus, by lemma 1.5.9 the invariants are given by

$$[T^q_{I,J}]^{\mathfrak{S}_{I \cap J}} = \left(p_I^* E^\vee \otimes (\wedge^{2c} N_I) \otimes p_J^* F\right)_{|\Delta_{I \cup J}} \otimes [\wedge^q N^\vee_{I \cap J | \Delta_{I \cup J}}]^{\mathfrak{S}_{I \cap J}} .$$

In particular, by the lemmas 1.5.20 and 1.5.22 the invariants vanish for q odd.

Remark 4.1.8. Let $k, p \in [n]$. We set $\mathcal{P}_p = \{J \subset [n] \mid \#J = p\}$. The orbits of $\mathcal{P}_p$ under the action of $\mathfrak{S}_{[k]} \times \mathfrak{S}_{\overline{[k]}}$ are exactly the sets ${}_k\mathcal{P}_p^i = \{J \in \mathcal{P}_p \mid \#(J \cap [k]) = i\}$ for $i = 0, \ldots, \min\{k, p\}$. A canonical choice of representatives of the orbits is

$${}_kI_p^i := \{1, \ldots, i, k+1, \ldots, k+p-i\} = [i] \cup [k+1, k+p-i] \in {}_k\mathcal{P}_p^i .$$

The stabiliser of ${}_kI_p^i$ is given by $\mathfrak{S}_{[k], {}_kI_p^i}$ (see above). Similarly, a system of representatives of the orbits of $\mathcal{P}_p$ under the $\mathfrak{S}_{[k]}$-action is given by all the sets of the form $[i] \cup M$ with $i = 0, \ldots, \min\{k, p\}$ and $M \subset [k+1, n]$.

Lemma 4.1.9. *Let E, F be locally free sheaves and B the spectral sequence described above. For $k = 0, \ldots, n-1$ the only non-vanishing term on the 2-sheet of $B(k)$ is $B(k)_2^{k,0}$.*

Proof. Using remark 4.1.2 and lemma 1.5.6 together with remark 1.5.8 yields

$$
\begin{aligned}
B(k)_1^{p,2k-q} &= [\mathcal{E}xt^{2k-q}(C_E^k, C_F^p)]^{\mathfrak{S}_n} \cong [\mathcal{E}xt^{2k-q}(\mathrm{Inf}^{\mathfrak{S}_n}_{\mathfrak{S}_{[k+1]} \times \overline{\mathfrak{S}_{[k+1]}}} E_{[k+1]}, C_F^p)]^{\mathfrak{S}_n} \\
&\cong [\mathcal{E}xt^{2k-q}(E_{[k+1]}, C_F^p)]^{\mathfrak{S}_{[k+1]} \times \overline{\mathfrak{S}_{[k+1]}}} \\
&\cong [\mathcal{E}xt^{2k-q}(E_{[k+1]}, \bigoplus_{I \in \mathcal{P}_{p+1}} F_I)]^{\mathfrak{S}_{[k+1]} \times \overline{\mathfrak{S}_{[k+1]}}} \qquad (1) \\
&\cong \bigoplus_{i=0}^{\min\{k,p\}} [\iota_{[k+1] \cup {}_kI_p^i *} T^q_{[k+1], {}_kI_p^i}]^{\mathfrak{S}_{[k+1], {}_kI_p^i}} . \qquad (2)
\end{aligned}
$$

By lemma 4.1.6 we see that $B(k)_1^{p,2k-q}$ vanishes whenever $p \notin \{k-1, k, k+1\}$. Furthermore, $B(k)_1^{p,2k-q}$ vanishes whenever q is odd by remark 4.1.7. Thus, the only non-trivial terms on the 1-level of $B(k)$ are organised in the short sequences

$$0 \to B(k)_1^{k-1,2k-q} \to B(k)_1^{k,2k-q} \to B(k)_1^{k+1,2k-q} \to 0$$

for $q \in [2k]$ even. We first will show that these sequences are exact for $q < 2k$, i.e. for $2k - q \geq 2$. By (1) they are isomorphic to the $(\mathfrak{S}_{k+1} \times \overline{\mathfrak{S}_{k+1}})$-invariants of the sequences

$$0 \to \mathcal{E}xt^{2k-q}(E_{[k+1]}, C_F^{k-1}) \to \mathcal{E}xt^{2k-q}(E_{[k+1]}, C_F^{k}) \to \mathcal{E}xt^{2k-q}(E_{[k+1]}, C_F^{k+1}) \to 0. \quad (3)$$

All sheaves in this sequences are push-forwards of sheaves on $\Delta_{[k+1]}$ so the $\mathfrak{S}_{k+1}$-linearization on them reduces to a $\mathfrak{S}_{k+1}$-action. We show that the sequence is already exact after taking the $\mathfrak{S}_{k+1}$-invariants. By (2) and lemma 4.1.6 the $\mathfrak{S}_{k+1}$-invariants of the sequence (3) are given by the sequence

$$[T_{[k+1],[k]}^q]^{\mathfrak{S}_k} \overset{\varphi}{\to} [T_{[k+1],[k+1]}^q]^{\mathfrak{S}_{k+1}} \oplus \bigoplus_{i=k+2}^{n} [T_{[k+1],[k]\cup\{i\}}^q]^{\mathfrak{S}_k} \overset{\psi}{\to} \bigoplus_{i=k+2}^{n} [T_{[k+1],[k+1]\cup\{i\}}^q]^{\mathfrak{S}_{k+1}} \quad (4)$$

where we left out the push-forwards along the closed embeddings in the notation. We denote the components of φ and ψ by

$$\varphi^0 \colon [T_{[k+1],[k]}^q]^{\mathfrak{S}_k} \to [T_{[k+1],[k+1]}^q]^{\mathfrak{S}_{k+1}}, \; \varphi^i \colon [T_{[k+1],[k]}^q]^{\mathfrak{S}_k} \to [T_{[k+1],[k]\cup\{i\}}^q]^{\mathfrak{S}_k}$$
$$\psi_0^j \colon [T_{[k+1],[k+1]}^q]^{\mathfrak{S}_{k+1}} \to [T_{[k+1],[k+1]\cup\{j\}}^q]^{\mathfrak{S}_{k+1}}, \; \psi_i^j \colon [T_{[k+1],[k]\cup\{i\}}^q]^{\mathfrak{S}_k} \to [T_{[k+1],[k+1]\cup\{j\}}^q]^{\mathfrak{S}_{k+1}}.$$

The direct summands occurring in (4) are of the form $[T_{I,J}^q]^{\mathfrak{S}_{I \cap J}}$ with $1 \in I \cap J$. Thus, by remark 4.1.7 they are given by

$$[\iota_{I \cup J *} T_{I,J}^q]^{\mathfrak{S}_{I \cap J}} = \iota_{I \cup J *}\left((p_1^* E^{\vee} \otimes (\wedge^{2c} N_I) \otimes p_1^* F)_{|\Delta_{I \cup J}} \otimes (\wedge^q \iota_{I \cup J}^* \mathcal{I}_{I \cap J})^{\mathfrak{S}_{I \cap J}} \right).$$

The differentials in $B(k)_1$ are induced by the differentials of the complex $C_F^\bullet$ whose components

$$\pi_{I,i} \colon F_I \cong \iota_{I *} \mathcal{O}_{\Delta_I} \otimes \mathrm{pr}_{\min I}^* F \to \iota_{I \cup \{i\} *} \mathcal{O}_{\Delta_{I \cup \{i\}}} \otimes \mathrm{pr}_{\min I}^* F \cong F_{I \cup \{i\}}$$

are given by the natural surjections $\iota_{I *} \mathcal{O}_{\Delta_I} \to \iota_{I \cup \{i\} *} \mathcal{O}_{\Delta_{I \cup \{i\}}}$ up to the sign $\varepsilon_{i, I \cup \{i\}}$. Thus, by 1.5.18 the differentials in the sequence (4) coincide up to the sign $\varepsilon_{i, I \cup \{i\}}$ with the maps induced by the inclusion $\mathcal{I}_{J \cap I} \subset \mathcal{I}_{J \cap (I \cup \{i\})}$ on the factor $(\wedge^q \iota_{I \cup J}^* \mathcal{I}_{I \cap J})^{\mathfrak{S}_{I \cap J}}$. A system of representatives of $\mathfrak{S}_{k+1}/\mathrm{Stab}_{\mathfrak{S}_{k+1}}([k])$ (see lemma 1.5.24) is given by the σ_ℓ of lemma 1.5.25. Thus, by Danila's lemma for morphisms (remark 1.5.7) the map φ^0 coincides with the morphism T from lemma 1.5.25 tensorized with the identity on $(p_1^* E^{\vee} \otimes \wedge^{2k} N_{[k+1]} \otimes p_1^* F)$. Thus, φ^0 is an isomorphism. This implies that φ is injective. Note that the ψ_i^j are zero if $i \neq 0$ and $i \neq j$. The morphisms ψ_j^j are isomorphisms by the same reason as φ^0 is. Thus, ψ is surjective. Moreover, we see that a local section in the kernel of ψ is determined by its component in $T_{[k+1],[k+1]}^q$. On the other hand, for every given local section s of $T_{[k+1],[k+1]}^q$ there is a section in the image of φ whose component in $T_{[k+1],[k+1]}^q$ equals s because of φ^0 being an isomorphism.

Since the rows in $B(k)_1$ are complexes, this already shows that $\mathrm{im}(\varphi) = \ker(\psi)$. So by now we have seen that all ℓ-th rows with $\ell > 0$ are exact on the 1-level what implies the vanishing of $B(k)_2^{p,\ell}$ for all $\ell > 0$ and all p. For $\ell = 2k - q = 0$, i.e. $q = 2k$, the $(\mathfrak{S}_{k+1} \times \overline{\mathfrak{S}_{k+1}})$-invariants of the sequence (3) reduce by (2) and remark 4.1.6 to the two term complex

$$0 \to [\mathcal{H}om(E_{[k+1]}, F_{[k+1]})]^{\mathfrak{S}_{[k+1]} \times \mathfrak{S}_{[k+2,n]}} \xrightarrow{\psi} [\mathcal{H}om(E_{[k+1]}, F_{[k+2]})]^{\mathfrak{S}_{[k+1]} \times \mathfrak{S}_{[k+3,n]}} \to 0 . \qquad (5)$$

The fact that the other terms vanish can be seen either directly by looking at the description of $T_{I,J}^{2k}$ or using the fact that for two sheaves $\mathcal{E}, \mathcal{F}$ on a variety which are push-forwards of locally free sheaves along closed immersions $\mathcal{H}om(\mathcal{E}, \mathcal{F})$ is non-trivial only if $\mathrm{supp}\,\mathcal{E} \supset \mathrm{supp}\,\mathcal{F}$. The morphism ψ is given (up to a sign) by composing morphisms $E_{[k+1]} \to F_{[k+1]}$ with the restriction homomorphism $F_{[k+1]} \to F_{[k+2]}$. Thus, before taking invariants it is surjective. The $\mathfrak{S}_{[k+1]}$ action on the domain and codomain is trivial (see remark 4.1.7). Thus, by corollary 3.7.6 the morphism of (5) is still surjective, which makes $B(k)_2^{0,k+1}$ vanish. The map ψ is not injective because the support of its domain is larger than the support of its codomain. So indeed, $B(k)_2^{k,0}$ is the only non-trivial term on the 2-level. $\qquad \square$

Corollary 4.1.10. *Let E, F be locally free sheaves on X and $k = 0 \ldots, n - 1$. Then the object $[R\,\mathcal{H}om(C_E^k, p_* q^* F^{[n]})]^{\mathfrak{S}_n}$ is cohomologically concentrated in degree k, i.e. for $m \neq k$*

$$[\mathcal{E}xt^m(C_E^k, p_* q^* F^{[n]})]^{\mathfrak{S}_n} = 0 .$$

Proof. The above lemma implies in particular that $B(k)$ degenerates at the 2-level. Thus, $B(k)_2^{p,q} = B(k)_\infty^{p,q}$ for all $p, q \in \mathbb{Z}$ and the only non-trivial term on the ∞-level is $B(k)_\infty^{k,0}$. Hence whenever $m \neq k$ we have

$$0 = B(k)^m = [\mathcal{E}xt^m(C_E^k, p_* q^* F^{[n]})]^{\mathfrak{S}_n} .$$

$\qquad \square$

Proposition 4.1.11. *Let E and F be locally free sheaves on X. Then $[R\,\mathcal{H}om(\Phi(E), \Phi(F))]^{\mathfrak{S}_n}$ is cohomogically concentrated in degree zero with*

$$[R\,\mathcal{H}om(\Phi(E^{[n]}), \Phi(F^{[n]}))]^{\mathfrak{S}_n} \simeq [\mathcal{H}om(C_E^0, C_F^0)]^{\mathfrak{S}_n} .$$

Proof. As mentioned above we use the spectral sequence

$$A_1^{p,q} = [\mathcal{E}xt^q(C_E^{-p}, p_* q^* F^{[n]})]^{\mathfrak{S}_n} \Longrightarrow A^m = [\mathcal{E}xt^m(C_E^\bullet, p_* q^* F^{[n]})]^{\mathfrak{S}_n} \simeq [\mathcal{E}xt^m(\Phi(E^{[n]}), \Phi(F^{[n]}))]^{\mathfrak{S}_n} .$$

By the above corollary the 1-sheet of A is concentrated on the diagonal $p + q = 0$. Thus,

$A^m = 0$ for $m \neq 0$. This yields

$$[R\,\mathcal{H}om(\Phi(E), \Phi(F))]^{\mathfrak{S}_n} \simeq [\mathcal{H}om(p_*q^*(E^{[n]}), p_*q^*(F^{[n]})))]^{\mathfrak{S}_n} \overset{4.1.1}{\simeq} [\mathcal{H}om(C_E^0, C_F^0)]^{\mathfrak{S}_n}.$$

$\square$

4.2 Generalisations

4.2.1 Determinant line bundles

Similar to what was done in section 3.8.3, we can generalise the propositions 4.1.4 and 4.1.11 to the case of tautological bundles twisted by determinant line bundles.

Proposition 4.2.1. *For locally free sheaves E, F and line bundles L, M on X the following holds*

(i) *The object $\left[R\,\mathcal{H}om(\Phi(E^{[n]} \otimes \mathcal{D}_L), \Phi(F^{[n]} \otimes \mathcal{D}_M)))\right]^{\mathfrak{S}_n}$ is cohomologically concentrated in degree zero and there is a natural isomorphism*

$$\left[\mathcal{H}om(p_*q^*(E^{[n]} \otimes \mathcal{D}_L), p_*q^*(F^{[n]} \otimes \mathcal{D}_M)))\right]^{\mathfrak{S}_n} \cong [\mathcal{H}om(C_E^0 \otimes L^{\boxtimes n}, C_F^0 \otimes M^{\boxtimes n})]^{\mathfrak{S}_n}.$$

(ii) *The object $\left[R\,\mathcal{H}om(\Phi(E^{[n]} \otimes \mathcal{D}_L), \Phi(\mathcal{D}_M))\right]^{\mathfrak{S}_n}$ is cohomologically concentrated in degree zero and there is a natural isomorphism*

$$\left[\mathcal{H}om(p_*q^*(E^{[n]} \otimes \mathcal{D}_L), p_*q^*(\mathcal{D}_M))\right]^{\mathfrak{S}_n} \cong [\mathcal{H}om(C_E^0 \otimes L^{\boxtimes n}, M^{\boxtimes n})]^{\mathfrak{S}_n}.$$

Proof. We will only show (i) since the proof of (ii) is very similar. Due to corollary 3.8.3 we have indeed

$$\left[R\,\mathcal{H}om(\Phi(E^{[n]} \otimes \mathcal{D}_L), \Phi(F^{[n]} \otimes \mathcal{D}_M)))\right]^{\mathfrak{S}_n} \simeq \left[R\,\mathcal{H}om(p_*q^*(E^{[n]} \otimes \mathcal{D}_L), p_*q^*(F^{[n]} \otimes \mathcal{D}_M)))\right]^{\mathfrak{S}_n}.$$

Furthermore,

$$\left[R\,\mathcal{H}om(\Phi(E^{[n]} \otimes \mathcal{D}_L), \Phi(F^{[n]} \otimes \mathcal{D}_M)))\right]^{\mathfrak{S}_n}$$
$$\overset{3.8.2}{\simeq} \left[R\,\mathcal{H}om(\Phi(E^{[n]}) \otimes L^{\boxtimes n}, \Phi(F^{[n]}) \otimes M^{\boxtimes n})\right]^{\mathfrak{S}_n}$$
$$\overset{\text{lf}}{\simeq} \left[\left(R\,\mathcal{H}om(\Phi(E^{[n]}), \Phi(F^{[n]})) \otimes (L^{\boxtimes n})^\vee \otimes M^{\boxtimes n})\right)\right]^{\mathfrak{S}_n}$$
$$\overset{3.8.2}{\simeq} [R\,\mathcal{H}om(\Phi(E^{[n]}), \Phi(F^{[n]})))]^{\mathfrak{S}_n} \otimes ((L^{\boxtimes n})^\vee)^{\mathfrak{S}_n} \otimes (M^{\boxtimes n})^{\mathfrak{S}_n}$$
$$\overset{4.1.11}{\simeq} [\mathcal{H}om(C_E^0, C_F^0)]^{\mathfrak{S}_n} \otimes ((L^{\boxtimes n})^\vee)^{\mathfrak{S}_n} \otimes (M^{\boxtimes n})^{\mathfrak{S}_n}$$

$$\overset{3.8.2}{\simeq} \left[\left(\mathcal{H}om(C_E^0, C_F^0) \otimes (L^{\boxtimes n})^{\vee} \otimes M^{\boxtimes n}\right)\right]^{\mathfrak{S}_n}$$
$$\overset{\text{lf}}{\simeq} [\mathcal{H}om(C_E^0 \otimes L^{\boxtimes n}, C_F^0 \otimes M^{\boxtimes n})]^{\mathfrak{S}_n} .$$

□

4.2.2 From tautological bundles to tautological objects

We call an object of the form $(E^{\bullet})^{[n]} \otimes \mathcal{D}_L \in \mathrm{D}^b(X^{[n]})$ with $E^{\bullet} \in \mathrm{D}^b(X)$ and L a line bundle on X a *twisted tautological object*.

Proposition 4.2.2. *Let $(E^{\bullet})^{[n]} \otimes \mathcal{D}_L$ and $(F^{\bullet})^{[n]} \otimes \mathcal{D}_M$ be twisted tautological objects. Then there are natural isomorphisms*

$$\left[R\mathcal{H}om(\Phi((E^{\bullet})^{[n]} \otimes \mathcal{D}_L), \Phi((F^{\bullet})^{[n]} \otimes \mathcal{D}_M))\right]^{\mathfrak{S}_n} \simeq \left[R\mathcal{H}om(C_{E^{\bullet}}^0 \otimes L^{\boxtimes n}, C_{F^{\bullet}}^0 \otimes M^{\boxtimes n})\right]^{\mathfrak{S}_n} ,$$

$$\left[R\mathcal{H}om((E^{\bullet})^{[n]} \otimes \mathcal{D}_L, \mathcal{D}_M)\right]^{\mathfrak{S}_n} \simeq \left[R\mathcal{H}om(C_{E^{\bullet}}^0 \otimes L^{\boxtimes n}, M^{\boxtimes n})\right]^{\mathfrak{S}_n} .$$

Proof. We take locally free resolutions $A^{\bullet} \simeq E^{\bullet}$ and $B^{\bullet} \simeq F^{\bullet}$ of the complexes $E^{\bullet}$ and $F^{\bullet}$. Then $(A^{\bullet})^{[n]} \simeq (E^{\bullet})^{[n]}$ and $(B^{\bullet})^{[n]} \simeq (F^{\bullet})^{[n]}$. Also, since

$$\Phi(E^{[n]} \otimes \mathcal{D}_L) \simeq Rp_*q^*(E^{[n]} \otimes \mathcal{D}_L) \simeq p_*q^*(E^{[n]} \otimes \mathcal{D}_L)$$

for every tautological sheaf $E^{[n]}$ and every line bundle $L \in \mathrm{Pic}(X)$ we have

$$\Phi((A^{\bullet})^{[n]} \otimes \mathcal{D}_L) \simeq p_*q^*((A^{\bullet})^{[n]} \otimes \mathcal{D}_L)$$

and similiar for $(B^{\bullet})^{[n]} \otimes \mathcal{D}_M$. Now for every pair $i, j \in \mathbb{Z}$ by proposition 4.2.1 we have

$$\underline{\mathrm{Ext}}^q(p_*q^*((A^i)^{[n]} \otimes \mathcal{D}_L), p_*q^*((B^j)^{[n]} \otimes \mathcal{D}_M)) = 0$$

for $q \neq 0$. Thus, we can apply proposition 1.5.1 to the situation of the bifunctor

$$\underline{\mathrm{Hom}} = [\_]^{\mathfrak{S}_n} \circ \pi_* \circ \mathcal{H}om(\_, \_) \colon \mathrm{QCoh}_{\mathfrak{S}_n}(X^n)^{\circ} \times \mathrm{QCoh}_{\mathfrak{S}_n}(X^n) \to \mathrm{QCoh}(S^n X)$$

and obtain using the naturalness of the isomorphism in 4.2.1 and the exactness of C^0

$$\left[R\mathcal{H}om(\Phi((E^{\bullet})^{[n]} \otimes \mathcal{D}_L), \Phi((F^{\bullet})^{[n]} \otimes \mathcal{D}_M))\right]^{\mathfrak{S}_n}$$
$$\simeq R\underline{\mathrm{Hom}}(p_*q^*((A^{\bullet})^{[n]} \otimes \mathcal{D}_L), p_*q^*((B^{\bullet})^{[n]} \otimes \mathcal{D}_M))$$
$$\simeq \underline{\mathrm{Hom}}^{\bullet}((p_*q^*((A^{\bullet})^{[n]} \otimes \mathcal{D}_L), p_*q^*((B^{\bullet})^{[n]} \otimes \mathcal{D}_M))$$

$$\simeq \left[\mathcal{H}om^\bullet(C^0_{A_\bullet} \otimes L^{\boxtimes n}, C^0_{B_\bullet} \otimes M^{\boxtimes n})\right]^{\mathfrak{S}_n}$$
$$\overset{\text{If}}{\simeq} \left[R\,\mathcal{H}om(C^0_{A_\bullet} \otimes L^{\boxtimes n}, C^0_{B_\bullet} \otimes M^{\boxtimes n})\right]^{\mathfrak{S}_n}$$
$$\simeq \left[R\,\mathcal{H}om(C^0_{E_\bullet} \otimes L^{\boxtimes n}, C^0_{F_\bullet} \otimes M^{\boxtimes n})\right]^{\mathfrak{S}_n}.$$

The second isomorphism is shown similarly. $\qquad\square$

4.3 Global Ext-groups

Theorem 4.3.1. *Let X be a smooth quasi-projective complex surface, $n \geq 2$, $E^\bullet, F^\bullet \in D^b(X)$, and $L, M \in \operatorname{Pic} X$. The extension groups of the associated twisted tautological objects are given by the following natural isomorphisms of graded vector spaces:*

$$\operatorname{Ext}^*((E^\bullet)^{[n]} \otimes \mathcal{D}_L, (F^\bullet)^{[n]} \otimes \mathcal{D}_M) \cong \begin{array}{l} \operatorname{Ext}^*(E^\bullet \otimes L, F^\bullet \otimes M) \otimes S^{n-1}\operatorname{Ext}^*(L, M)\oplus \\ \operatorname{Ext}^*(E^\bullet \otimes L, M) \otimes \operatorname{Ext}^*(L, F^\bullet \otimes M) \otimes S^{n-2}\operatorname{Ext}^*(L, M), \end{array}$$
$$\operatorname{Ext}^*((E^\bullet)^{[n]} \otimes \mathcal{D}_L, \mathcal{D}_M) \cong \operatorname{Ext}^*(E^\bullet \otimes L, M) \otimes S^{n-1}\operatorname{Ext}^*(L, M).$$

Proof. Using the previous proposition and the considerations at the beginning of this chapter, the extension groups are given by

$$\operatorname{Ext}^* \left((E^\bullet)^{[n]} \otimes \mathcal{D}_L, (F^\bullet)^{[n]} \otimes \mathcal{D}_M\right)$$
$$\cong \mathfrak{S}_n \operatorname{Ext}^* \left(\Phi((E^\bullet)^{[n]} \otimes \mathcal{D}_L), \Phi((F^\bullet)^{[n]} \otimes \mathcal{D}_M)\right)$$
$$\cong \operatorname{H}^*(S^n X, \left[R\,\mathcal{H}om(\Phi((E^\bullet)^{[n]} \otimes \mathcal{D}_L), \Phi((F^\bullet)^{[n]} \otimes \mathcal{D}_M))\right]^{\mathfrak{S}_n})$$
$$\cong \operatorname{H}^*(S^n X, \left[R\,\mathcal{H}om(C^0_{E_\bullet} \otimes L^{\boxtimes n}, C^0_{F_\bullet} \otimes M^{\boxtimes n})\right]^{\mathfrak{S}_n})$$
$$\cong \left[\operatorname{H}^*(X^n, R\,\mathcal{H}om(C^0_{E_\bullet} \otimes L^{\boxtimes n}, C^0_{F_\bullet} \otimes M^{\boxtimes n}))\right]^{\mathfrak{S}_n}.$$

Applying the adjoint property of the inflation functor for $C^0_{E_\bullet} \otimes L^{\boxtimes} \simeq \operatorname{Inf}^{\mathfrak{S}_n}_{\mathfrak{S}_{[1]}}(p_1^* E \otimes L^{\boxtimes n})$ and Danilas lemma we get

$$\left[R\,\mathcal{H}om(C^0_{E_\bullet} \otimes L^{\boxtimes n}, C^0_{F_\bullet} \otimes M^{\boxtimes n})\right]^{\mathfrak{S}_n}$$
$$\simeq \left[\bigoplus_{j \in [n]} R\,\mathcal{H}om(p_1^* E \otimes L^{\boxtimes n}, p_j^* F \otimes M^{\boxtimes n})\right]^{\overline{\mathfrak{S}_{[1]}}}$$
$$\simeq \left[R\,\mathcal{H}om(p_1^* E^\bullet \otimes L^{\boxtimes n}, p_1^* F^\bullet \otimes M^{\boxtimes n})\right]^{\overline{\mathfrak{S}_{[1]}}} \oplus \left[R\,\mathcal{H}om(p_1^* E^\bullet \otimes L^{\boxtimes n}, p_2^* F^\bullet \otimes M^{\boxtimes n})\right]^{\overline{\mathfrak{S}_{[2]}}}. \quad (*)$$

Using the compatibility of the derived sheaf-Hom with pull-backs gives

$$R\mathcal{H}om(p_1^*E^\bullet \otimes L^{\boxtimes n}, p_1^*F^\bullet \otimes M^{\boxtimes n}) \simeq R\mathcal{H}om(E^\bullet \otimes L, F^\bullet \otimes M) \boxtimes \mathcal{H}om(L, M)^{\boxtimes n-1}$$

and

$$R\mathcal{H}om(p_1^*E^\bullet \otimes L^{\boxtimes n}, p_2^*F^\bullet \otimes M^{\boxtimes n})$$
$$\simeq R\mathcal{H}om(E^\bullet \otimes L, M) \boxtimes \mathcal{H}om(L, F^\bullet \otimes M) \boxtimes \mathcal{H}om(L, M)^{\boxtimes n-2}.$$

Now by the Künneth formula

$$\left[\mathrm{H}^*\left(X^n, R\mathcal{H}om(E^\bullet \otimes L, F^\bullet \otimes M) \boxtimes \mathcal{H}om(L, M)^{\boxtimes n-1}\right) \right]^{\overline{\mathfrak{S}_{[1]}}}$$
$$\cong \left[\mathrm{H}^*(R\mathcal{H}om(E^\bullet \otimes L, F^\bullet \otimes M)) \otimes \mathrm{H}^*(\mathcal{H}om(L, M))^{\otimes n-1} \right]^{\overline{\mathfrak{S}_{[1]}}}$$
$$\cong \left[\mathrm{Ext}^*(E^\bullet \otimes L, F^\bullet \otimes M)) \otimes \mathrm{Ext}^*(L, M)^{\otimes n-1} \right]^{\overline{\mathfrak{S}_{[1]}}}$$
$$\cong \mathrm{Ext}^*(E^\bullet \otimes L, F^\bullet \otimes M)) \otimes S^{n-1}(\mathrm{Ext}^*(L, M)).$$

Doing the same for the other direct summand in $(*)$ yields the result. The proof of the second formula is again similar but easier and therefore omitted. $\qquad\square$

Remark 4.3.2. The above formulas are natural in $E^\bullet$, $F^\bullet$, L, and M, in automorphisms of X, and also in pull-backs along open immersions $U \subset X$.

Remark 4.3.3. By setting $L = M = \mathcal{O}_X$ we get the following formulas for non-twisted tautological objects

$$\mathrm{Ext}^*((E^\bullet)^{[n]}, (F^\bullet)^{[n]}) \cong \begin{array}{l} \mathrm{Ext}^*(E^\bullet, F^\bullet) \otimes S^{n-1}(\mathrm{H}^*(\mathcal{O}_X)) \oplus \\ \mathrm{H}^*((E^\bullet)^\vee) \otimes \mathrm{H}^*(E^\bullet) \otimes S^{n-2}(\mathrm{H}^*(\mathcal{O}_X)) \end{array}$$
$$\mathrm{Ext}^*((E^\bullet)^{[n]}, \mathcal{O}_X) \cong \mathrm{H}^*(X^{[n]}, ((E^\bullet)^{[n]})^\vee) \cong \mathrm{H}^*((E^\bullet)^\vee) \otimes S^{n-1}\mathrm{H}^*(\mathcal{O}_X).$$

Remark 4.3.4. By the same arguments as used in this and the previous subsection we can also generalise Scala's formula for the cohomology of twisted tautological sheaves (see theorem 2.5.5) to twisted tautological objects. Namely, for every tautological object $(E^\bullet)^{[n]} \otimes \mathcal{D}_L$ there is a natural isomorphism of graded vector spaces

$$\mathrm{H}^*(X^{[n]}, (E^\bullet)^{[n]} \otimes \mathcal{D}_L) \cong \mathrm{H}^*(E^\bullet \otimes L) \otimes S^{n-1}\mathrm{H}^*(L).$$

Also, since $\mathcal{D}_L^\vee \otimes \mathcal{D}_M \cong \mathcal{D}_{\mathcal{H}om(L,M)}$ for $L, M \in \mathrm{Pic}(X)$, we have a formula for

$$\mathrm{Ext}^*(\mathcal{D}_L, (E^\bullet)^{[n]} \otimes \mathcal{D}_M) \cong \mathrm{H}^*(X^{[n]}, (E^\bullet)^{[n]} \otimes \mathcal{D}_{\mathcal{H}om(L,M)}).$$

Remark 4.3.5. In the case that X is projective one can also directly deduce the formula for $\text{Ext}^*((E^\bullet)^{[n]} \otimes \mathcal{D}_L, \mathcal{D}_M)$ by Serre duality from Scala's formula in the form of the previous remark and the fact that $\mathcal{D}_{\omega_X} = \omega_{X^{[n]}}$.

Remark 4.3.6. Using proposition 4.1.1 we also get for $E_1, \ldots, E_k, F$ locally free sheaves on X formulas for $\text{Ext}^0(E_1^{[n]} \otimes \cdots \otimes E_k^{[n]}, \mathcal{O}_{X^{[n]}})$ as well as for $\text{Ext}^0(E_1^{[n]} \otimes \cdots \otimes E_k^{[n]}, F^{[n]})$. For $\ell \in \mathbb{N}$ and $M \subset \mathbb{N}$ let $P(M, \ell)$ we denote the set of partitions $I = \{I_1, \ldots, I_\ell\}$ of the set M of lenght ℓ. Then

$$\text{Hom}(E_1^{[n]} \otimes \cdots \otimes E_k^{[n]}, \mathcal{O}_{X^{[n]}}) \cong \bigoplus_{I \in P([k], \ell), \ell \leq n} \left(\bigotimes_{s=1}^{\ell} \text{H}^0(\otimes_{i \in I_s} E_i^\vee) \otimes S^{n-\ell} \text{H}^0(\mathcal{O}_X) \right)$$

and $\text{Hom}(E_1^{[n]} \otimes \cdots \otimes E_k^{[n]}, F^{[n]})$ is isomorphic to

$$\bigoplus_{\substack{M \subset [k] \\ I \in P([k] \backslash M, \ell), \ell \leq n-1}} \left(\text{Hom}(\otimes_{i \in M} E_i, F) \bigotimes_{s=1}^{\ell} \text{H}^*(\otimes_{i \in I_s} E_i^\vee) \otimes S^{n-\ell-1} \text{H}^0(\mathcal{O}_X) \right).$$

Again, there are similar formulas for the sheaves twisted by determinant line bundles. However, these formulas can not directly be generalised to formulas for Ext^* since the corresponding $R\underline{\text{Hom}}$ objects are in general not cohomologically concentrated in degree zero for $k \geq 2$. Note that for X projective the first formula (generalised to tensor products of tautological bundles twisted by determinant line bundles) is via Serre duality the same as proposition 3.8.6.

4.4   Spherical and $\mathbb{P}^n$-objects

Let X be a smooth projective variety with canonical bundle ω_X. We call an object $\mathcal{E}^\bullet \in \text{D}^b(X)$ an ω_X-*invariant* object if $\mathcal{E}^\bullet \otimes \omega_X \simeq \mathcal{E}^\bullet$. A *spherical object* in $\text{D}^b(X)$ is an ω_X-invariant object $\mathcal{E}^\bullet$ with the property

$$\text{Ext}^i(\mathcal{E}^\bullet, \mathcal{E}^\bullet) = \begin{cases} \mathbb{C} & \text{if } i = 0, \dim(X), \\ 0 & \text{else,} \end{cases}$$

i.e. $\text{Ext}^*(\mathcal{E}^\bullet, \mathcal{E}^\bullet) \cong \text{H}^*(S^{\dim X}, \mathbb{C})$, where S^n denotes the real n-sphere. A $\mathbb{P}^n$-*object* in the derived category $\text{D}^b(X)$ is a ω_X-invariant object $\mathcal{F}^\bullet$ such that there is an isomorphism of graded algebras $\text{Ext}^*(\mathcal{F}^\bullet, \mathcal{F}^\bullet) \cong \text{H}^*(\mathbb{P}^n, \mathbb{C})$, where the multiplication on the left is given by the Yoneda product and on the right by the cup product. In particular for the underlying vector spaces

$$\operatorname{Ext}^i(\mathcal{E}^\bullet, \mathcal{E}^\bullet) = \begin{cases} \mathbb{C} & \text{if } 0 \leq i \leq 2n \text{ is even} \\ 0 & \text{if } i \text{ is odd} \end{cases}$$

holds. By Serre duality the dimension of X must be $2n$ as soon as $D^b(X)$ contains a $\mathbb{P}^n$-object. Spherical and $\mathbb{P}^n$-objects are of interest because they induce automorphisms of $D^b(X)$ (see [Huy06, chapter 8]). For a smooth projective surface X with $\omega_X = \mathcal{O}_X$ the canonical bundle on $X^{[n]}$ is also trivial (see remark 3.8.2). Hence the property of being $\omega_{X^{[n]}}$-invariant is automatically satisfied for every object in $D^b(X^{[n]})$. Thus, one could hope that there are tautological objects induced by some special objects in $D^b(X)$ that are spherical or $\mathbb{P}^n$-objects. But this is not the case by the following proposition.

Proposition 4.4.1. *(i) Let X be a smooth projective surface with trivial canonical line bundle and $n \geq 2$. Then twisted tautological objects on $X^{[n]}$ are never spherical or $\mathbb{P}^n$-objects.*

(ii) Let X be a smooth projective surface and $n \geq 3$. Then twisted tautological sheaves on $X^{[n]}$ are never spherical or $\mathbb{P}^n$-objects.

Proof. Let $(E^\bullet)^{[n]} \otimes \mathcal{D}_L$ be a non-zero twisted tautological object on $X^{[n]}$. Then using the fact that $\mathcal{H}om(L, L) \cong \mathcal{O}_X$ we have

$$\operatorname{Ext}^*((E^\bullet)^{[n]} \otimes \mathcal{D}_L, (E^\bullet)^{[n]} \otimes \mathcal{D}_L) \cong \operatorname{Ext}^*((E^\bullet)^{[n]}, (E^\bullet)^{[n]}).$$

Thus, we may assume $L = \mathcal{O}_X$. The graded vector space $\operatorname{Ext}^*((E^\bullet)^{[n]}, (E^\bullet)^{[n]})$ is given by theorem 4.3.1 by

$$\operatorname{Ext}^*(E^\bullet, E^\bullet) \otimes S^{n-1} \operatorname{H}^*(\mathcal{O}_X) \oplus \operatorname{H}^*((E^\bullet)^\vee) \otimes \operatorname{H}^*(E^\bullet) \otimes S^{n-2} \operatorname{H}^*(\mathcal{O}_X).$$

If $\omega_X = \mathcal{O}_X$ we have by Serre duality $h^2(\mathcal{O}_X) = h^0(\mathcal{O}_X) = 1$. The vector space $\operatorname{Ext}^0(E^\bullet, E^\bullet)$ has positive dimension since it contains the identity. Thus $\operatorname{Ext}^0(E^\bullet, E^\bullet) \otimes S^{n-1} \operatorname{H}^*(\mathcal{O}_X)$ contributes non-trivially to the degrees $0, 2, \ldots, 2n - 2$ of $\operatorname{Ext}^*((E^\bullet)^{[n]}, (E^\bullet)^{[n]})$. But by Serre duality also $\operatorname{Ext}^2(E^\bullet, E^\bullet)$ is non-vanishing. Thus, also $\operatorname{Ext}^2(E^\bullet, E^\bullet) \otimes S^{n-1} \operatorname{H}^*(\mathcal{O}_X)$ contributes non-trivially to the degrees $2, 4, \ldots, 2n$. This yields $\operatorname{ext}^i((E^\bullet)^{[n]}, (E^\bullet)^{[n]}) \geq 2$ for $i = 2, 4, \ldots, 2n - 2$ which shows that $(E^\bullet)^{[n]} \otimes \mathcal{D}_L$ is indeed neither a spherical nor a $\mathbb{P}^n$-object. Now let $n \geq 3$ and $E^{[n]}$ be a tautological sheaf. Extension groups of sheaves on X can be non-trivial only in the degrees $0, 1, 2$. If $E^{[n]}$ was a spherical or a $\mathbb{P}^n$-object, then in particular the highest and the lowest extension groups, i.e. in degree 0 and $2n$, must not vanish. Since for $n \geq 3$ the term $\operatorname{H}^*(\mathcal{O}_X)$ occurs in both direct summands of $\operatorname{Ext}^*((E^\bullet)^{[n]}, (E^\bullet)^{[n]})$ it follows that $\operatorname{H}^i(\mathcal{O}_X) \neq 0$ for $i = 0, 2$. Furthermore, either $\operatorname{Ext}^2(E, E)$

or $(\text{Ext}^*(E, \mathcal{O}_X) \otimes \text{Ext}^*(\mathcal{O}_X, E))^4$ must not vanish. In both cases $\text{ext}^4(E^{[n]}, E^{[n]}) \geq 2$ since also $\text{Ext}^0(E, E) \otimes S^{n-1} \text{H}^*(\mathcal{O}_X)$ contributes non-trivially to $\text{Ext}^4(E^{[n]}, E^{[n]})$. $\qquad \square$

4.5 Products and interpretation of the results

4.5.1 Yoneda products, the Künneth isomorphism and signs

Let $\mathcal{A}$ be an abelian category with enough injectives. Recall that for every $A^\bullet \in \text{D}^-(\mathcal{A})$ and $B^\bullet \in \text{D}^+(\mathcal{A})$ there are natural isomorphisms $\text{Ext}^i(A^\bullet, B^\bullet) \cong \text{Hom}_{\text{D}(\mathcal{A})}(A^\bullet, B^\bullet[i])$. For $A^\bullet, B^\bullet, C^\bullet \in \text{D}^b(\mathcal{A})$ the *Yoneda product* is the bilinear map Yon making the following diagram commute

$$
\begin{array}{ccccc}
\text{Ext}^j(B^\bullet, C^\bullet) & \times & \text{Ext}^i(A^\bullet, B^\bullet) & \xrightarrow{\ \text{Yon}\ } & \text{Ext}^{i+j}(A^\bullet, C^\bullet) \\
\downarrow{\cong} & & \downarrow{\cong} & & \downarrow{\cong} \\
\text{Hom}_{\text{D}(\mathcal{A})}(B^\bullet[i], C^\bullet[i+j]) & \times & \text{Hom}_{\text{D}(\mathcal{A})}(A^\bullet, B^\bullet[i]) & \xrightarrow{\ \circ\ } & \text{Hom}_{\text{D}(\mathcal{A})}(A^\bullet, C^\bullet[i+j]).
\end{array}
$$

The second and third vertical arrows are the isomorphisms mentioned and the first is the mentioned isomorphism composed with the i-th power of the shift functor of $\text{D}(\mathcal{A})$. The lower horizontal bilinear map is just the composition law in $\text{D}(\mathcal{A})$. In the following we will treat the vertical isomorphisms as they were identities and consequently denote the Yoneda product just by $\circ$. Considering the above products as maps on the homogeneous components we get the Yoneda product as a bilinear map of graded vector spaces

$$
\text{Ext}^*(B^\bullet, C^\bullet) \times \text{Ext}^*(A^\bullet, B^\bullet) \xrightarrow{\circ} \text{Ext}^*(A^\bullet, C^\bullet).
$$

There will occur signs because of the use of the Künneth isomorphism. Let X, Y be varieties together with objects $A_1^\bullet, B_1^\bullet, C_1^\bullet \in \text{D}^b(X)$ and $A_2^\bullet, B_2^\bullet, C_2^\bullet \in \text{D}^b(Y)$. We consider homogeneous elements $a_i \in \text{Ext}^*(A_i^\bullet, B_i^\bullet)$ and $b_i \in \text{Ext}^*(B_i^\bullet, C_i^\bullet)$ for $i = 1, 2$. Then via the Künneth isomorphism we can interpret the tensor products as elements of the extension groups on the product $X \times Y$ namely

$$
a_1 \otimes a_2 \in \text{Ext}^*(A_1^\bullet \boxtimes A_2^\bullet, B_1^\bullet \boxtimes B_2^\bullet), \ b_1 \otimes b_2 \in \text{Ext}^*(B_1^\bullet \boxtimes B_2^\bullet, C_1^\bullet \boxtimes C_2^\bullet).
$$

The Yoneda product is then given by

$$
(b_1 \otimes b_2) \circ (a_1 \otimes a_2) = (-1)^{\deg b_2 \deg a_1}(b_1 a_1) \otimes (b_2 a_2),
$$

where we omit the $\circ$ for the Yoneda products on X and Y. The occurrence of the sign can be seen best when defining the Yoneda product using the cup product (see e.g [HL10, section 10.1.1]). To capture these signs we use the following convention: Let $a_1, \ldots, a_n$ and $b_1 = a_{n+1}, \ldots, b_n = a_{2n}$ be elements in any graded algebraic objects of degree $\deg(a_i) = p_i$. Let T be a term in which all the a_i and b_i occur. Let $\sigma \in \mathfrak{S}_{2n}$ be the permutation such that after erasing everything in T besides the a_i and b_i we get $a_{\sigma^{-1}(1)} \ldots a_{\sigma^{-1}(2n)}$. Then we set $\varepsilon(T) := \varepsilon_{\sigma, p_1, \ldots, p_{2n}}$ (see section 1.3). Let I be a finite set and for each $i \in I$ let T_i be a term in which all the a_i and b_i occur. We define

$$\sum_{i \in I}^{\bullet} T_i := \sum_{i \in I} \varepsilon(T_i) \cdot T_i \,.$$

In the following we will have such sums where a and b will be replaced by two other letters. Then always the first letter from the left in each summand will be the same. This letter is considered as a and the other as b. For example if we have x_1, x_2, y_1, y_2 all of odd degree then

$$\sum_{\sigma, \tau \in \mathfrak{S}_2}^{\bullet} (x_{\sigma^{-1}(1)} \circ y_{\tau^{-1}(1)}) \otimes (x_{\sigma^{-1}(2)} \circ y_{\tau^{-1}(2)})$$
$$= -(x_1 \circ y_1) \otimes (x_2 \circ y_2) + (x_1 \circ y_2) \otimes (x_2 \circ y_1) + (x_2 \circ y_1) \otimes (x_1 \circ y_2) - (x_2 \circ y_2) \otimes (x_1 \circ y_1) \,.$$

4.5.2 Yoneda products for twisted tautological objects

Let $E^\bullet, F^\bullet \in D^b(X)$ and L, M be line bundles on X. In the last section we computed formulas for $\operatorname{Ext}^*((E^\bullet)^{[n]} \otimes \mathcal{D}_L, \mathcal{D}_M)$ as well as for $\operatorname{Ext}^*((E^\bullet)^{[n]} \otimes \mathcal{D}_L, (F^\bullet)^{[n]} \otimes \mathcal{D}_M)$. It was done using natural isomorphisms

$$\operatorname{Ext}^*((E^\bullet)^{[n]} \otimes \mathcal{D}_L, \mathcal{D}_M) \cong \mathfrak{S}_n \operatorname{Ext}^*(\Phi((E^\bullet)^{[n]} \otimes \mathcal{D}_L), \Phi(\mathcal{D}_M)) \cong \mathfrak{S}_n \operatorname{Ext}^*(C^0_{E^\bullet} \otimes L^{\boxtimes n}, M^{\boxtimes n}) \tag{1}$$

respectively

$$\operatorname{Ext}^*((E^\bullet)^{[n]} \otimes \mathcal{D}_L, (F^\bullet)^{[n]} \otimes \mathcal{D}_M) \cong \mathfrak{S}_n \operatorname{Ext}^*(\Phi((E^\bullet)^{[n]} \otimes \mathcal{D}_L), \Phi((F^\bullet)^{[n]} \otimes \mathcal{D}_M)) \tag{2}$$
$$\cong \mathfrak{S}_n \operatorname{Ext}^*(C^0_{E^\bullet} \otimes L^{\boxtimes n}, C^0_{F^\bullet} \otimes M^{\boxtimes n}) \,.$$

There is also a formula for $\operatorname{Ext}^*(\mathcal{D}_L, (F^\bullet)^{[n]} \otimes \mathcal{D}_M)$ (see remark 4.3.4) using the isomorphisms

$$\operatorname{Ext}^*(\mathcal{D}_L, (F^\bullet)^{[n]} \otimes \mathcal{D}_M) \cong \mathfrak{S}_n \operatorname{Ext}^*(\Phi(\mathcal{D}_L), \Phi((F^\bullet)^{[n]} \otimes \mathcal{D}_M)) \cong \mathfrak{S}_n \operatorname{Ext}^*(L^{\boxtimes n}, C^0_{F^\bullet} \otimes M^{\boxtimes n}) \,. \tag{3}$$

Furthermore, the formula

$$\operatorname{Ext}^*(\mathcal{D}_L, \mathcal{D}_M) \simeq S^n \operatorname{Ext}^*(L, M) \tag{4}$$

can easily be proven using the isomorphisms (see lemma 3.8.2)

$$\mathrm{Ext}^*(\mathcal{D}_L, \mathcal{D}_M) \cong \mathfrak{S}_n \, \mathrm{Ext}^*(\Phi(\mathcal{D}_L), \Phi(\mathcal{D}_M)) \cong \mathfrak{S}_n \, \mathrm{Ext}^*(L^{\boxtimes n}, M^{\boxtimes n}) \,. \tag{5}$$

In summary, we have formulas for the extension groups $\mathrm{Ext}^*(\mathcal{E}^\bullet, \mathcal{F}^\bullet)$ on $X^{[n]}$ whenever each of $\mathcal{E}^\bullet$ and $\mathcal{F}^\bullet$ is a twisted tautological object or a determinant line bundle. Clearly the Yoneda products

$$\mathrm{Ext}^*(\mathcal{F}^\bullet, \mathcal{G}^\bullet) \times \mathrm{Ext}^*(\mathcal{E}^\bullet, \mathcal{F}^\bullet) \to \mathrm{Ext}^*(\mathcal{E}^\bullet, \mathcal{G}^\bullet)$$

coincide under the Bridgeland-King-Reid equivalence with the Yoneda products

$$\mathfrak{S}_n \, \mathrm{Ext}^*(\Phi(\mathcal{F}^\bullet), \Phi(\mathcal{G}^\bullet)) \times \mathfrak{S}_n \, \mathrm{Ext}^*(\Phi(\mathcal{E}^\bullet), \Phi(\mathcal{F}^\bullet)) \to \mathfrak{S}_n \, \mathrm{Ext}^*(\Phi(\mathcal{E}^\bullet), \Phi(\mathcal{G}^\bullet))$$

because Φ is a functor. Now for locally free sheaves A and B on X the isomorphism

$$[\mathcal{H}om(C_A^0, C_B^0)]^{\mathfrak{S}_n} \xrightarrow{\cong} [\mathcal{H}om(p_*q^*(A^{[n]}), p_*q^*(B^{[n]}))]^{\mathfrak{S}_n}$$

is given by restricting the $\mathfrak{S}_n$-equivariant morphisms to the subsheaf $p_*q^*(A^{[n]})$ of C_A^0 and observing that the restricted morphisms factorise through the subsheaf $p_*q^*(B^{[n]})$ of C_B^0 (see proposition 4.1.1). Also the isomorphisms $[\mathcal{H}om(C_A^0, \mathcal{O}_{X^n})]^{\mathfrak{S}_n} \xrightarrow{\cong} [\mathcal{H}om(p_*q^*(A^{[n]}), \mathcal{O}_{X^n})]^{\mathfrak{S}_n}$ and $[\mathcal{H}om(\mathcal{O}_{X^n}, C_B^0)]^{\mathfrak{S}_n} \xrightarrow{\cong} [\mathcal{H}om(\mathcal{O}_{X^n}, p_*q^*(B^{[n]}))]^{\mathfrak{S}_n}$ are given by restriction of the morphisms to subsheaves. Now the isomorphisms on the right hand sides of (1), (2) and (3) are induced by those isomorphisms of the sheaf-Homs after choosing locally free resolutions of $E^\bullet$ and $F^\bullet$. The composition of morphisms of sheaves is compatible with restricting the morphisms to certain subsheaves. This translates into the Yoneda products between the extension groups of twisted tautological objects and determinant line bundles coinciding with the Yoneda products between the equivariant extension groups of the terms of the form $C_{E^\bullet}^0$, $L^{\boxtimes n}$, and $C_{E^\bullet}^0 \otimes L^{\boxtimes n}$ under the isomorphisms in (1), (2), (3) and (5). Now using the Künneth and Danila's isomorphism we can express the Yoneda products in terms of Yoneda products in the derived category of the surface X under the isomorphisms of theorem 4.3.1, remark 4.3.4 and (4) . We will explicitly state and prove the formula only for the case were all the objects $\mathcal{E}^\bullet$, $\mathcal{F}^\bullet$ and $\mathcal{G}^\bullet$ involved are twisted tautological objects. The other seven cases can be done very similarly. For $E, F \in \mathrm{D}^b(X)$ and $L, M \in \mathrm{Pic}(X)$ we set

$$P(E, L, F, M) := \begin{array}{l} \mathrm{Ext}^*(E \otimes L, F \otimes M) \otimes S^{n-1}(\mathrm{Ext}^*(L, M)) \oplus \\ \mathrm{Ext}^*(E \otimes L, M) \otimes \mathrm{Ext}^*(L, F \otimes M) \otimes S^{n-2}(\mathrm{Ext}^*(L, M)) \,. \end{array}$$

Let also be $G \in \mathrm{D}^b(X)$ and $N \in \mathrm{Pic}(X)$. We consider the elements

$$\begin{pmatrix} \varphi \otimes s_2 \cdots s_n \\ \eta \otimes x \otimes t_3 \cdots t_n \end{pmatrix} \in P(F, M, G, N) \cong \mathrm{Ext}^*(F^{[n]} \otimes \mathcal{D}_M, G^{[n]} \otimes \mathcal{D}_N)$$

$$\begin{pmatrix} \varphi' \otimes s_2' \cdots s_n' \\ \eta' \otimes x' \otimes t_3' \cdots t_n' \end{pmatrix} \in P(E, L, F, M) \cong \mathrm{Ext}^*(E^{[n]} \otimes \mathcal{D}_L, F^{[n]} \otimes \mathcal{D}_M).$$

In order to use the sign convention above we set

$$\varphi = s_1 \,, \; \eta = t_1 \,, \; x = t_2 \,, \; \varphi' = s_1' \,, \; \eta' = t_1' \,, \; x' = t_2' \,.$$

and assume that all the s_i, s_i', t_i and t_i' are homogeneous. Now we can compute the Yoneda product $\begin{pmatrix} \varphi \otimes s_2 \cdots s_n \\ \eta \otimes x \otimes t_3 \cdots t_n \end{pmatrix} \circ \begin{pmatrix} \varphi' \otimes s_2' \cdots s_n' \\ \eta' \otimes x' \otimes t_3' \cdots t_n' \end{pmatrix}$ in $\mathrm{Ext}^*(E^{[n]} \otimes \mathcal{D}_L, G^{[n]} \otimes \mathcal{D}_N)$ and express it as an element in $P(E, L, G, M)$.

Proposition 4.5.1. *The Yoneda product is given by*

$$\frac{1}{(n-1)!} \sum_{\sigma \in \mathfrak{S}_{[2,n]}}^{\bullet} (\varphi\varphi') \otimes (s_2 s_{\sigma^{-1}(2)}') \cdots (s_n s_{\sigma^{-1}(n)}')$$

$$+ \frac{n-1}{(n-2)!} \sum_{\tau \in \mathfrak{S}_{[3,n]}}^{\bullet} (x\eta') \otimes (\eta x') \cdot (t_3 t_{\tau^{-1}(3)}') \cdots (t_n t_{\tau^{-1}(n)}')$$

$$\oplus \frac{1}{(n-1)!} \sum_{\beta \in \mathfrak{S}_{[2,n]}}^{\bullet} (\eta\varphi') \otimes (x s_{\beta^{-1}(2)}') \otimes (t_3 s_{\beta^{-1}(3)}') \cdots (t_n s_{\beta^{-1}(n)}')$$

$$+ \frac{1}{(n-1)!} \sum_{\gamma \in \mathfrak{S}_{[2,n]}}^{\bullet} (s_{\gamma^{-1}(2)}\eta') \otimes (\varphi x') \otimes (s_{\gamma^{-1}(3)}t_3') \cdots (s_{\tau^{-1}(n)}t_n')$$

$$+ \frac{1}{(n-2)!} \sum_{\substack{i=3,\ldots,n \\ \alpha \in \mathfrak{S}_{[3,\ldots,n]}}}^{\bullet} (t_i\eta') \otimes (x t_{\alpha^{-1}(i)}') \otimes (\eta x') \cdot (t_3 t_{\alpha^{-1}(3)}') \cdots \widehat{(t_i t_{\alpha^{-1}(i)}')} \cdots (t_n t_{\alpha^{-1}(n)}') \,.$$

Proof. The element $\begin{pmatrix} \varphi \otimes s_2 \cdots s_n \\ \eta \otimes x \otimes t_3 \cdots t_n \end{pmatrix} \in P(F, M, G, N)$ corresponds to the element

$$\frac{1}{(n-1)!} \sum_{\sigma \in \mathfrak{S}_n}^{\bullet} s_{\sigma^{-1}(1)} \otimes \cdots \otimes s_{\sigma^{-1}(n)} \oplus \frac{1}{(n-2)!} \sum_{\tau \in \mathfrak{S}_n}^{\bullet} t_{\tau^{-1}(1)} \otimes \cdots \otimes t_{\tau^{-1}(n)}$$

in $\mathrm{Ext}^*(C_F^0 \otimes M^{\boxtimes n}, C_G^0 \otimes N^{\boxtimes n})$. The coefficients are coming from the canonical isomorphism $S^k V \xrightarrow{\cong} S_k V$ (see section 1.3, note that the isomorphism of Danila's lemma does not involve such coefficients). The same holds for $\begin{pmatrix} \varphi' \otimes s_2' \cdots s_n' \\ \eta' \otimes x' \otimes t_3' \cdots t_n' \end{pmatrix}$. The summand $s_{\sigma^{-1}(1)} \otimes \cdots \otimes s_{\sigma^{-1}(n)}$ is an element of $\mathrm{Ext}^*(\mathrm{pr}_{\sigma(1)}^* F \otimes M^{\boxtimes n}, \mathrm{pr}_{\sigma(1)}^* G \otimes N^{\boxtimes n})$ and $t_{\tau^{-1}(1)} \otimes \cdots \otimes t_{\tau^{-1}(2)}$ an element of $\mathrm{Ext}^*(\mathrm{pr}_{\tau(1)}^* F \otimes M^{\boxtimes n}, \mathrm{pr}_{\tau(2)}^* G \otimes N^{\boxtimes n})$. There are five types of composable pairs of the components of the classes in $\mathrm{Ext}^*(C_F^0 \otimes M^{\boxtimes n}, C_G^0 \otimes N^{\boxtimes n}) \times \mathrm{Ext}^*(C_E^0 \otimes L^{\boxtimes n}, C_F^0 \otimes M^{\boxtimes n})$, namely

$$\mathrm{pr}_i^* E \otimes L^{\boxtimes n} \to \mathrm{pr}_i^* F \otimes M^{\boxtimes n} \to \mathrm{pr}_i^* G \otimes N^{\boxtimes n} \;,\; \mathrm{pr}_i^* E \otimes L^{\boxtimes n} \to \mathrm{pr}_j^* F \otimes M^{\boxtimes n} \to \mathrm{pr}_i^* G \otimes N^{\boxtimes n},$$

$$\mathrm{pr}_i^* E \otimes L^{\boxtimes n} \to \mathrm{pr}_i^* F \otimes M^{\boxtimes n} \to \mathrm{pr}_j^* G \otimes N^{\boxtimes n} \;,\; \mathrm{pr}_i^* E \otimes L^{\boxtimes n} \to \mathrm{pr}_j^* F \otimes M^{\boxtimes n} \to \mathrm{pr}_j^* G \otimes N^{\boxtimes n},$$

$$\mathrm{pr}_i^* E \otimes L^{\boxtimes n} \to \mathrm{pr}_j^* F \otimes M^{\boxtimes n} \to \mathrm{pr}_k^* G \otimes N^{\boxtimes n}$$

with $i, j, k \in [n]$ pairwise distinct. Thus, the Yoneda product in $\mathrm{Ext}^*(C_E^0 \otimes L^{\boxtimes n}, C_G^0 \otimes N^{\boxtimes n})$ looks like this :

$$\frac{1}{(n-1)!^2} \sum_{\substack{\sigma,\sigma' \in \mathfrak{S}_n \\ \sigma(1)=\sigma'(1)}}^{\bullet} \otimes_{i=1}^n (s_{\sigma^{-1}(i)} s'_{\sigma'^{-1}(i)}) + \frac{1}{(n-2)!^2} \sum_{\substack{\tau,\tau' \in \mathfrak{S}_n \\ \tau(2)=\tau'(1),\tau(1)=\tau'(2)}}^{\bullet} \otimes_{i=1}^n (t_{\tau^{-1}(i)} t'_{\tau'^{-1}(i)})$$

$$+ \frac{1}{(n-1)!(n-2)!} \sum_{\substack{\tau,\sigma' \in \mathfrak{S}_n \\ \tau(1)=\sigma'(1)}}^{\bullet} \otimes_{i=1}^n (t_{\tau^{-1}(i)} s'_{\sigma'^{-1}(i)}) + \frac{1}{(n-1)!(n-2)!} \sum_{\substack{\sigma,\tau' \in \mathfrak{S}_n \\ \sigma(1)=\tau'(2)}}^{\bullet} \otimes_{i=1}^n (s_{\sigma^{-1}(i)} t'_{\tau'^{-1}(i)})$$

$$+ \frac{1}{(n-2)!^2} \sum_{\substack{\tau,\tau' \in \mathfrak{S}_n \\ \tau(1)=\tau'(2),\tau(2)\neq\tau'(1)}}^{\bullet} \otimes_{i=1}^n (t_{\tau^{-1}(i)} t'_{\tau'^{-1}(i)}) \quad . \tag{6}$$

The first term of (6) is a $\mathfrak{S}_n$-invariant element of $\oplus_{i=1}^n \mathrm{Ext}^*(p_i^* E \otimes L^{\boxtimes n}, p_i^* G \otimes N^{\boxtimes n})$. Danila's isomorphism is simply the projection to the first summand. Thus, it maps the first term of (6) to

$$\frac{1}{(n-1)!^2} \sum_{\substack{\sigma,\sigma' \in \mathfrak{S}_n \\ \sigma(1)=\sigma'(1)=1}}^{\bullet} \otimes_{i=1}^n (s_{\sigma^{-1}(i)} s'_{\sigma'^{-1}(i)}) \in \mathrm{Ext}^*(p_1^* E \otimes L^{\boxtimes n}, p_1^* G \otimes N^{\boxtimes n})$$

Under the isomorphism $S_{n-1} \mathrm{Ext}^*(L, N) \xrightarrow{\cong} S^{n-1} \mathrm{Ext}^*(L, N)$ this element is mapped to

$$\frac{1}{(n-1)!} \sum_{\sigma \in \mathfrak{S}_{[2,n]}}^{\bullet} (\varphi\varphi') \otimes (s_2 s'_{\sigma^{-1}(2)}) \cdots (s_n s'_{\sigma^{-1}(n)}) \in \mathrm{Ext}^*(E \otimes L, G \otimes N) \otimes S^{n-1} \mathrm{Ext}^*(L, N)$$

which is exactly the first term of the formula we want to prove. Doing the same for the other four terms in (6) yields the desired element in $P(E, L, G, M)$. $\square$

Remark 4.5.2. Let $D, E, F, G \in \mathrm{D}^b(X)$ and $L, M, N \in \mathrm{Pic}\, X$. By remark 2.5.4 for a morphism $\varphi \in \mathrm{Hom}_{\mathrm{D}^b(X)}(E, F[i]) \cong \mathrm{Ext}^i(E, F)$ the induced morphism

$$\varphi^{[n]} \otimes \mathrm{id}_{\mathcal{D}_M} \in \mathrm{Ext}^i(E^{[n]} \otimes \mathcal{D}_M, F^{[n]} \otimes \mathcal{D}_M)$$

corresponds to

$$\begin{pmatrix} \varphi \otimes \mathrm{id}_M \cdots \mathrm{id}_M \\ 0 \end{pmatrix} \in P(E, M, F, M).$$

For $M = \mathcal{O}_X$ this shows that the tautological functor $(\_)^{[n]}$ is faithful but not full. Let

$$\begin{pmatrix} \psi' \otimes s_2' \cdots s_n' \\ \eta' \otimes x' \otimes t_3' \cdots t_n' \end{pmatrix} \in P(D, L, E, M), \quad \begin{pmatrix} \psi \otimes s_2 \cdots s_n \\ \eta \otimes x \otimes t_3 \cdots t_n \end{pmatrix} \in P(F, M, G, N).$$

Then the formula for the Yoneda product gives back the naturalness of the isomorphism in theorem 4.3.1 as a special case since

$$\begin{pmatrix} \psi \otimes s_2 \cdots s_n \\ \eta \otimes x \otimes t_3 \cdots t_n \end{pmatrix} \circ \begin{pmatrix} \varphi \otimes \mathrm{id}_M \cdots \mathrm{id}_M \\ 0 \end{pmatrix} = \begin{pmatrix} \psi\varphi \otimes s_2 \cdots s_n \\ \eta\varphi \otimes x \otimes t_3 \cdots t_n \end{pmatrix}$$

$$\begin{pmatrix} \varphi \otimes \mathrm{id}_M \cdots \mathrm{id}_M \\ 0 \end{pmatrix} \circ \begin{pmatrix} \psi' \otimes s_2' \cdots s_n' \\ \eta' \otimes x' \otimes t_3' \cdots t_n' \end{pmatrix} = \begin{pmatrix} \varphi\psi' \otimes s_2' \cdots s_n' \\ \eta' \otimes \varphi x' \otimes t_3' \cdots t_n' \end{pmatrix}.$$

4.5.3   Interpretation of the formulas

Let $W \subset S^n X$ be the closed subset of unordered tuples $x_1 + \cdots + x_n$ with $|\{x_1, \ldots, x_n\}| < n$, i.e. $W = \pi(\mathbb{D})$ where $\mathbb{D}$ denotes the big diagonal in X^n. We denote the complement open subset by $S^n X_* = S^n X \setminus W$. Analogous to what is done in subsection 3.1.2 we set $X_*^n := \pi^{-1}(S^n X_*) = X^n \setminus \mathbb{D}$, $X_*^{[n]} := \mu^{-1}(S^n D_*)$ and $I^n X_* = q^{-1}(X_*^{[n]}) = p^{-1}(X_*^n)$. We also set $\Xi_* = \mathrm{pr}_{X_*^{[n]}}^{-1} X_*^{[n]}$, $Z_* = \mathrm{pr}_{I^n X}^{-1} I^n X_* = \Xi_* \times_{X_*^{[n]}} I^n X_*$, and $D_* = \mathrm{pr}_{X_*^n}^{-1} X_*^n$. Again, we will write $(\_)_*$ for the restricted morphisms and sheaves. The morphism $\mu_* \colon X_*^{[n]} \to S^n X_*$ and $p_* \colon I^n X_* \to X_*^n$ are isomorphisms. The varieties Ξ_* and D_* are given by the n-fold disjoint unions of $X_*^{[n]} \cong S^n X_*$ respectively $D_{i*} \cong X_*^n$. Furthermore, the group $\mathfrak{S}_n$ acts freely on X_*^n and $I^n X_*$. Thus, the quotient morphisms π_* and q_* are flat. Under the identifications $X_*^{[n]} \cong S^n X_*$ and $I^n X_* \cong X_*^n$ the Bridgeland–King–Reid equivalence is given simply as the pull-back functor $(\pi_*)^* \colon \mathrm{D}^b(X_*^{[n]}) \to \mathrm{D}^b_{\mathfrak{S}_n}(X_*^n)$. For every $F \in \mathrm{Coh}(X)$ the isomorphism $\Phi_*(F_*^{[n]}) \simeq \Phi_{D_*}(F)$ is simply given by the flat base change

$$\begin{array}{ccc} D_* = Z_* & \longrightarrow & \Xi_* & \longrightarrow & X \\ \downarrow & & \downarrow & & \\ X_*^n & \xrightarrow{\;\pi_*\;} & S^n X_*. & & \end{array}$$

The augmentation map $\hat{\gamma}_* \colon \mathcal{O}_{D_*} \to \oplus_{i=1}^n \mathcal{O}_{D_{i*}} = \mathcal{K}_*^0$ of $\mathcal{O}_{D_*}$ into the resolution $\mathcal{K}_*^\bullet$ (see section 2.5) is an isomorphism. Thus, the morphism $\gamma_* \colon \Phi_{\mathcal{O}_{D_*}}(F) \simeq (\pi_*)^* F^{[n]} \to C_F^0$ is an isomorphism. For a locally free sheaf the $F \in \mathrm{Coh}(X)$ the fibers of the tautological sheaf $F^{[n]}$ over closed points $[\xi]$ are given by $F^{[n]}([\xi]) = \Gamma(\xi, F_{|\xi})$. If $[\xi] \in X_*^{[n]}$, i.e. ξ is of the form $\xi = \{x_1, \ldots, x_n\}$, the fiber is given by the direct sum of the $F(x_i)$. More generally, for a line bundle L on X we have

$$(F^{[n]} \otimes \mathcal{D}_L)([\xi]) = (F \otimes L)(x_1) \oplus \cdots \oplus (F \otimes L)(x_n)$$

By the above discussion and applying the inverse of the isomorphism of Danila's lemma, we see that in Scala's formula for cohomological degree zero

$$\mathrm{H}^0(X^{[n]}, F^{[n]} \otimes \mathcal{D}_L) \cong \mathrm{H}^0(X^n, C_F^0 \otimes L^{\boxtimes n}) \cong \mathrm{H}^0(F \otimes L) \otimes S^{n-1}\mathrm{H}^0(L)$$

an element $f \otimes s_2 \cdots s_n$ of the right-hand side, i.e. $f \in \mathrm{H}^0(F \otimes L)$ and $s_2, \ldots, s_n \in \mathrm{H}^0(L)$, corresponds to the global section $S \in \mathrm{H}^0(X^{[n]}, F^{[n]} \otimes \mathcal{D}_L)$ given over $X_*^{[n]}$ by

$$S([\xi])(x_i) = \frac{1}{(n-1)!} f(x_i) \cdot \left(\sum_{\substack{\sigma \in \mathfrak{S}_n \\ \sigma(1)=i}} \prod_{\ell \in [n] \setminus \{i\}} s_{\sigma^{-1}(\ell)}(x_\ell) \right).$$

Here we have to choose local trivialisations inducing isomorphisms $L(x_i) \cong \mathbb{C}$ in order to interpret the $s_j(x_i)$ as elements of $\mathbb{C}$. Since sections of locally free sheaves on varieties are determined by their values on open sets, the assignment $S \leftrightarrow f \otimes s_2 \cdots s_n$ gives the isomorphism of theorem 2.5.5 in an explicit form in degree 0. For general X the Hilbert scheme $X^{[n]}$ is covered by open subsets of the form $U_i^{[n]}$ with $U_i \subset X$ affine (see lemma 3.7.1). Let $\mathcal{U}^{[n]} = \{U_i^{[n]} \mid i \in I\}$ be such a cover induced by an appropriate open affine cover $\mathcal{U} = \{U_i \mid i \in I\}$ of X. The intersections of these open sets are given by

$$U_{i_0, \ldots, i_n} := U_{i_0}^{[n]} \cap \cdots \cap U_{i_n}^{[n]} = (U_{i_0} \cap \cdots \cap U_{i_n})^{[n]}.$$

By Scala's formula 2.5.5 the higher cohomology of $F^{[n]} \otimes \mathcal{D}_L$ restricted to these open subsets vanishes. Thus, $\mathcal{U}^{[n]}$ is a Čech cover for $F^{[n]} \otimes \mathcal{D}_L$, i.e.

$$\mathrm{H}^*(X^{[n]}, F^{[n]} \otimes \mathcal{D}_L) = \check{H}(\mathcal{U}^{[n]}, F^{[n]} \otimes \mathcal{D}_L).$$

For $S^n \mathcal{U} = \{S^n U_i \mid i \in I\}$ also

$$\mathrm{H}^*(S^n X, [C_F^0 \otimes L^{\boxtimes n}]^{\mathfrak{S}_n}) = \check{H}(S^n \mathcal{U}, [C_F^0 \otimes L^{\boxtimes n}]^{\mathfrak{S}_n}).$$

Thus, the isomorphism of theorem 2.5.5 in terms of Čech cohomology is given on every open set the same way as it is in degree zero. More concretely, the cohomology class $f \otimes s_2 \cdots s_n$ with $s_1 := f \in \mathrm{H}^{q_1}(F \otimes L)$ and $s_i \in \mathrm{H}^{q_i}(L)$ such that $q_1 + \cdots + q_n = q$ corresponds to

$S \in \mathrm{H}^q(X^{[n]}, F^{[n]} \otimes \mathcal{D}_L)$ given for $\xi = \{x_1, \ldots, x_n\}$ with $[\xi] \in U^{[n]}_{i_0,\ldots,i_{q*}}$ by

$$S_{i_0,\ldots,i_q}([\xi])(x_i) = \frac{1}{(n-1)!} \cdot \left(\sum_{\substack{\sigma \in \mathfrak{S}_n \\ \sigma(1)=i}}^{\bullet} (s_{\sigma^{-1}(1)})_{i_0,\ldots,i_{q_{\sigma^{-1}(1)}}}(x_1) \cdots (s_{\sigma^{-1}(n)})_{i_{q-q_{\sigma^{-1}(n)}},\ldots,i_q}(x_n) \right).$$

For $F^\bullet \in \mathrm{D}^b(X)$ we choose a locally free resolution $A^\bullet$ of $F^\bullet$. Then also $(A^\bullet)^{[n]}$ is a locally free resolution of $(F^\bullet)^{[n]}$. Thus, $\mathrm{H}^*(X^{[n]}, (F^\bullet)^{[n]} \otimes \mathcal{D}_L)$ is computed by the Čech hypercohomology with values in $(A^\bullet)^{[n]} \otimes \mathcal{D}_L$. Hence, we get the same describtion as above of the isomorphism

$$\mathrm{H}^*(X^{[n]}, (F^\bullet)^{[n]} \otimes \mathcal{D}_L) \cong \mathrm{H}^*(F^\bullet \otimes L) \otimes S^{n-1}\, \mathrm{H}^*(L)$$

in terms of Čech hypercohomology. Similarly, we can interpret theorem 4.3.1, i.e. the formula

$$\mathrm{Ext}^* \left((E^\bullet)^{[n]} \otimes \mathcal{D}_L, (F^\bullet)^{[n]} \otimes \mathcal{D}_M \right) \cong P(E^\bullet, L, F^\bullet, M).$$

For this, note that

$$\mathrm{Ext}^* \left((E^\bullet)^{[n]} \otimes \mathcal{D}_L, (F^\bullet)^{[n]} \otimes \mathcal{D}_M \right) \cong \mathrm{H}^* \left(X^{[n]}, R\,\mathcal{H}om((E^\bullet)^{[n]} \otimes \mathcal{D}_L, (F^\bullet)^{[n]} \otimes \mathcal{D}_M) \right).$$

Let E, F be locally free sheaves and L, M line bundles on X. Then an element

$$\begin{pmatrix} \psi \otimes s_2 \cdots s_n \\ \eta \otimes f \otimes t_3 \cdots t_n \end{pmatrix} \in P(E, L, F, M)$$

of cohomological degree 0 corresponds to the morphism $\Psi \colon E^{[n]} \otimes \mathcal{D}_L \to F^{[n]} \otimes \mathcal{D}_M$ given as follows. For $\xi = \{x_1, \ldots, x_n\}$ we denote by $\Psi([\xi]) \colon E^{[n]} \to F^{[n]}$ the restriction to the fiber and by $\Psi([\xi])(i,j) \colon F(x_i) \to F(x_j)$ for $i, j \in [n]$ the components of this map. Then

$$F([\xi])(i,i) = \frac{1}{(n-1)!} \psi(x_i) \cdot \left(\sum_{\substack{\sigma \in \mathfrak{S}_n \\ \sigma(1)=i}} \prod_{\ell \in [n] \setminus \{i\}} s_{\sigma^{-1}(\ell)}(x_\ell) \right) \tag{1}$$

and for $i \neq j$ the component is given by

$$\Psi([\xi])(i,j) = \frac{1}{(n-2)!} f(x_j) \circ \eta(x_i) \cdot \left(\sum_{\substack{\sigma \in \mathfrak{S}_n \\ \sigma(1)=i, \sigma(2)=j}} \prod_{\ell \in [n] \setminus \{i,j\}} t_{\sigma^{-1}(\ell)}(x_\ell) \right).$$

Since $\mathcal{U}^{[n]}$ is a Čech cover for the sheaf $\mathcal{H}om(E^{[n]}{\otimes}\mathcal{D}_L, F^{[n]}{\otimes}\mathcal{D}_M)$ we can make the same generalisations as above to get an explicit description of the isomorphism of theorem 4.3.1 in terms of Čech hypercohomology. Furthermore, the isomorphism $\mathrm{H}^0(X^{[n]}, \mathcal{O}_{X^{[n]}}) \cong S^n\,\mathrm{H}^0(\mathcal{O}_X)$ is given by sending $s_1 \cdots s_n \in H^0(\mathcal{O}_X)$ to the global regular function S on $X^{[n]}$ given over $X_*^{[n]}$ by

$$S([\xi]) = \frac{1}{n!} \sum_{\sigma \in \mathfrak{S}_n} s_{\sigma^{-1}(1)}(x_1) \cdots s_{\sigma^{-1}(n)}(x_n)\,. \tag{2}$$

Also this isomorphism can be generalised directly to higher cohomological degrees.

4.5.4 The trace map and the cup product

Proposition 4.5.3. *For $E^\bullet \in \mathrm{D}^b(X)$ and L a line bundle on X the global trace map*

$$\mathrm{tr}_{(E^\bullet)^{[n]}\otimes\mathcal{D}_L} \colon\; \mathrm{Ext}^*((E^\bullet)^{[n]} \otimes \mathcal{D}_L, (E^\bullet)^{[n]} \otimes \mathcal{D}_L) \to \mathrm{H}^*(X^{[n]}, \mathcal{O}_{X^{[n]}})$$

is given under the isomorphisms $\mathrm{Ext}^*((E^\bullet)^{[n]} \otimes \mathcal{D}_L, (E^\bullet)^{[n]} \otimes \mathcal{D}_L) \cong P(E^\bullet, L, E^\bullet, L)$ *and* $\mathrm{H}^*(X^{[n]}, \mathcal{O}_{X^{[n]}}) \cong S^n\,\mathrm{H}^*(\mathcal{O}_X)$ *by*

$$\begin{pmatrix} \psi \otimes s_2 \cdots s_n \\ \eta \otimes f \otimes t_3 \cdots t_n \end{pmatrix} \mapsto n \cdot (\mathrm{tr}_{E^\bullet \otimes L}(\varphi) \cdot s_2 \cdots s_n)\,.$$

Proof. Since the sheaf trace map is computed by taking a locally free resolution and then computing the trace map in every degree, we can assume that $E^\bullet \simeq E$ is a locally free sheaf. Since the global trace map is obtained in Čech cohomology by applying the trace map on every open set, we can by the above discussion restrict the proof to the case of cohomological degree zero. So let

$$\begin{pmatrix} \psi \otimes s_2 \cdots s_n \\ \eta \otimes f \otimes t_3 \cdots t_n \end{pmatrix} \in P(E, L, E, L)$$

be an element of degree zero corresponding to the endomorphism Ψ of $E^{[n]} \otimes \mathcal{D}_L$. The image under the trace map can be computed fiberwise and is determined by its value over $X_*^{[n]}$. The trace of Ψ on the fiber over $\xi = \{x_1, \ldots x_n\}$ is indeed given by

$$\begin{aligned}
\mathrm{tr}\big(\Psi([\xi])\big) &= \sum_{i=1}^n \mathrm{tr}\big(\Psi([\xi])(i,i)\big) \\
&\overset{(1)}{=} \frac{1}{(n-1)!} \sum_{i=1}^n \sum_{\substack{\sigma \in \mathfrak{S}_n \\ \sigma(1)=i}} \mathrm{tr}(\psi)(x_i) \cdot \prod_{\ell \in [n]\setminus\{i\}} s_{\sigma^{-1}(\ell)}(x_\ell) \\
&= \frac{1}{(n-1)!} \sum_{\sigma \in \mathfrak{S}_n} \mathrm{tr}(\psi)(x_{\sigma(1)}) \cdot s_2(x_{\sigma(2)}) \cdots s_n(x_{\sigma(n)}) \overset{(2)}{=} n \cdot S([\xi])
\end{aligned}$$

122

with S being the global regular function on $X^{[n]}$ corresponding to $\mathrm{tr}(\psi)\cdot s_2\cdots s_n \in S^n\,\mathrm{H}^0(\mathcal{O}_X)$.

$\square$

Remark 4.5.4. Let E, F be locally free sheaves on X. By theorem 3.7.9 the cohomology of $E^{[n]}\otimes F^{[n]}$ is isomorphic to a graded subspace of $\mathrm{H}^*(X^n, C_E^0\otimes C_F^0)$ which in turn is isomorphic to

$$Q(E,F) := \mathrm{H}^*(E\otimes F)\otimes S^{n-1}\,\mathrm{H}^*(\mathcal{O}_X)\oplus \mathrm{H}^*(E)\otimes \mathrm{H}^*(F)\otimes S^{n-2}\,\mathrm{H}^*(\mathcal{O}_X)\,.$$

Again we see that the higher cohomology vanishes on $U^{[n]}$ for U affine, i.e. $\mathcal{U}^{[n]}$ is a Čech cover for $E^{[n]}\otimes F^{[n]}$. Thus, as before a description of the isomorphisms in degree zero yields a description in every degree. For $[\xi]\in X_*^{[n]}$ as above the fiber of $E^{[n]}\otimes F^{[n]}$ is given by

$$(E^{[n]}\otimes F^{[n]})([\xi]) = \bigoplus_{i,j\in[n]} E(x_i)\otimes F(x_j)\,.$$

An element

$$\begin{pmatrix} a\otimes s_2\cdots s_n \\ e\otimes f\otimes t_3\cdots t_n \end{pmatrix}\in Q(E,F)$$

corresponds to the global section S of $E^{[n]}\otimes F^{[n]}$ which is given over $X_*^{[n]}$ by

$$S([\xi])(i,i) = \frac{1}{(n-1)!}a(x_i)\cdot\left(\sum_{\substack{\sigma\in\mathfrak{S}_n \\ \sigma(1)=i}}\prod_{\ell\in[n]\setminus\{i\}} s_{\sigma^{-1}(\ell)}(x_\ell)\right)\in E(x_i)\otimes F(x_i)$$

and for $i\neq j$ by

$$S([\xi])(i,j) = \frac{1}{(n-2)!}e(x_i)\otimes f(x_j)\cdot\left(\sum_{\substack{\sigma\in\mathfrak{S}_n \\ \sigma(1)=i,\sigma(2)=j}}\prod_{\ell\in[n]\setminus\{i,j\}} t_{\sigma^{-1}(\ell)}(x_\ell)\right)\in E(x_i)\otimes F(x_j)\,.$$

Using this one can compute that the cup product

$$\cup\colon \mathrm{H}^*(X^{[n]}, E^{[n]})\times \mathrm{H}^*(X^{[n]}, F^{[n]})\to \mathrm{H}^*(X^{[n]}, E^{[n]}\otimes F^{[n]})$$

of $e\otimes s_2\cdots s_n\in\mathrm{H}^*(E)\otimes S^{n-1}\mathrm{H}(\mathcal{O}_X)$ and $f\otimes t_2\cdots t_n\in\mathrm{H}^*(F)\otimes S^{n-1}\mathrm{H}(\mathcal{O}_X)$ is given by

$$\begin{pmatrix} \frac{1}{(n-1)!}\sum_{\sigma\in\mathfrak{S}_{[2,n]}}^\bullet (e\cup f)\otimes\prod_{\ell=2}^n(s_\ell\cup t_{\sigma^{-1}(\ell)}) \\ \frac{1}{(n-1)(n-1)!}\sum_{\substack{i\in[2,n] \\ \sigma\in\mathfrak{S}_{[2,n]}}}^\bullet (e\cup t_{\sigma^{-1}(i)})\otimes(s_i\cup f)\otimes\prod_{\ell\in[2,n]\setminus\{i\}}(s_\ell\cup t_{\sigma^{-1}(\ell)}) \end{pmatrix}\,.$$

Here in order to use the sign convention of subsection 4.5.1 we have to set $e = s_1$ as well as $f = t_1$. More generally, one can compute that for $E_1,\ldots,E_k$ locally free sheaves the composition of the k-fold cup product with the map induced on cohomology by the inclusion of K_k into K_0, i.e.

$$\bigotimes_{i=1}^{k} \mathrm{H}(X^{[n]}, E_i^{[n]}) \xrightarrow{\cup} \mathrm{H}(X^{[n]}, \bigotimes_{i=1}^{k} E_i^{[n]}) \cong \mathrm{H}^*(X^n, K_k)^{\mathfrak{S}_n} \to \mathrm{H}^*(X^n, K_0)^{\mathfrak{S}_n}$$

coincides with the map Ψ from section 3.3.

Bibliography

[BKR01]   Tom Bridgeland, Alastair King, and Miles Reid. The McKay correspondence as an equivalence of derived categories. *J. Amer. Math. Soc.*, 14(3):535–554 (electronic), 2001.

[BNW07]   Samuel Boissière and Marc A. Nieper-Wißkirchen. Universal formulas for characteristic classes on the Hilbert schemes of points on surfaces. *J. Algebra*, 315(2):924–953, 2007.

[Dan00]   Gentiana Danila. Sections du fibré déterminant sur l'espace de modules des faisceaux semi-stables de rang 2 sur le plan projectif. *Ann. Inst. Fourier (Grenoble)*, 50(5):1323–1374, 2000.

[Dan01]   Gentiana Danila. Sur la cohomologie d'un fibré tautologique sur le schéma de Hilbert d'une surface. *J. Algebraic Geom.*, 10(2):247–280, 2001.

[Dan07]   Gentiana Danila. Sections de la puissance tensorielle du fibré tautologique sur le schéma de Hilbert des points d'une surface. *Bull. Lond. Math. Soc.*, 39(2):311–316, 2007.

[DN89]   J.-M. Drezet and M. S. Narasimhan. Groupe de Picard des variétés de modules de fibrés semi-stables sur les courbes algébriques. *Invent. Math.*, 97(1):53–94, 1989.

[FH91]   William Fulton and Joe Harris. *Representation theory*, volume 129 of *Graduate Texts in Mathematics*. Springer-Verlag, New York, 1991. A first course, Readings in Mathematics.

[Fog68]   John Fogarty. Algebraic families on an algebraic surface. *Amer. J. Math*, 90:511–521, 1968.

[Ful98]   William Fulton. *Intersection theory*, volume 2 of *Ergebnisse der Mathematik und ihrer Grenzgebiete. 3. Folge. A Series of Modern Surveys in Mathematics [Results in Mathematics and Related Areas. 3rd Series. A Series of Modern Surveys in Mathematics]*. Springer-Verlag, Berlin, second edition, 1998.

[GM96]   Sergei I. Gelfand and Yuri I. Manin. *Methods of homological algebra*. Springer-Verlag, Berlin, 1996. Translated from the 1988 Russian original.

[Gro57]   Alexander Grothendieck. Sur quelques points d'algèbre homologique. *Tôhoku Math. J. (2)*, 9:119–221, 1957.

[Gro95]   Alexander Grothendieck. Techniques de construction et théorèmes d'existence en géométrie algébrique. IV. Les schémas de Hilbert. In *Séminaire Bourbaki, Vol. 6*, pages Exp. No. 221, 249–276. Soc. Math. France, Paris, 1995.

[Hai01]   Mark Haiman. Hilbert schemes, polygraphs and the Macdonald positivity conjecture. *J. Amer. Math. Soc.*, 14(4):941–1006 (electronic), 2001.

[Hai02]   Mark Haiman. Vanishing theorems and character formulas for the Hilbert scheme of points in the plane. *Invent. Math.*, 149(2):371–407, 2002.

[Har66]   Robin Hartshorne. *Residues and duality*. Lecture notes of a seminar on the work of A. Grothendieck, given at Harvard 1963/64. With an appendix by P. Deligne. Lecture Notes in Mathematics, No. 20. Springer-Verlag, Berlin, 1966.

[Har77]   Robin Hartshorne. *Algebraic geometry*. Springer-Verlag, New York, 1977. Graduate Texts in Mathematics, No. 52.

[HL10]   Daniel Huybrechts and Manfred Lehn. *The geometry of moduli spaces of sheaves*. Cambridge Mathematical Library. Cambridge University Press, Cambridge, second edition, 2010.

[Huy06]   D. Huybrechts. *Fourier-Mukai transforms in algebraic geometry*. Oxford Mathematical Monographs. The Clarendon Press Oxford University Press, Oxford, 2006.

[KS06]   Masaki Kashiwara and Pierre Schapira. *Categories and sheaves*, volume 332 of *Grundlehren der Mathematischen Wissenschaften [Fundamental Principles of Mathematical Sciences]*. Springer-Verlag, Berlin, 2006.

[Leh99]   Manfred Lehn. Chern classes of tautological sheaves on Hilbert schemes of points on surfaces. *Invent. Math.*, 136(1):157–207, 1999.

[LH09]   Joseph Lipman and Mitsuyasu Hashimoto. *Foundations of Grothendieck duality for diagrams of schemes*, volume 1960 of *Lecture Notes in Mathematics*. Springer-Verlag, Berlin, 2009.

[Nak01]   Iku Nakamura. Hilbert schemes of abelian group orbits. *J. Algebraic Geom.*, 10(4):757–779, 2001.

[NW04]   Marc Nieper-Wißkirchen. *Chern numbers and Rozansky-Witten invariants of compact hyper-Kähler manifolds*. World Scientific Publishing Co. Inc., River Edge, NJ, 2004.

[Sca09a]   Luca Scala. Cohomology of the Hilbert scheme of points on a surface with values in representations of tautological bundles. *Duke Math. J.*, 150(2):211–267, 2009.

[Sca09b]   Luca Scala. Some remarks on tautological sheaves on Hilbert schemes of points on a surface. *Geom. Dedicata*, 139:313–329, 2009.

[Ser00]   Jean-Pierre Serre. *Local algebra*. Springer Monographs in Mathematics. Springer-Verlag, Berlin, 2000. Translated from the French by CheeWhye Chin and revised by the author.

Andreas Krug

Universität Augsburg
Lehrstuhl für Algebra und Zahlentheorie
E-Mail: andreas_krug@yahoo.de

Persönliche Daten

Geburtstag:	27.12.1983
Geburtsort:	Rüsselsheim

Hochschulbildung

April 2009-August 2012	Promotionsstudium Mathematik Universität Augsburg
Oktober 2006-März 2009	Hauptstudium Mathematik Universität Mainz
Oktober 2004-September 2006	Grundstudium Mathematik Universität Mainz

Schulbildung

Sommer 1994-Sommer 2003	Max-Planck Gymnasium Rüsselsheim
Sommer 1990-Sommer 1994	Eichgrundschule Rüsselsheim

Beruflicher Werdegang

April 2009-August 2012	Wissenschaftlicher Mitarbeiter Universität Augsburg
Oktober 2006- März 2008	Studentische Hilfskraft Universität Mainz
Oktober 2003- Juli 2004	Zivildienst Wicherngemeinde Rüsselsheim